PRIVACY PRESERVING
DATA MINING

Advances in Information Security

Sushil Jajodia
Consulting Editor
Center for Secure Information Systems
George Mason University
Fairfax, VA 22030-4444
email: jajodia@gmu.edu

The goals of the Springer International Series on ADVANCES IN INFORMATION SECURITY are, one, to establish the state of the art of, and set the course for future research in information security and, two, to serve as a central reference source for advanced and timely topics in information security research and development. The scope of this series includes all aspects of computer and network security and related areas such as fault tolerance and software assurance.

ADVANCES IN INFORMATION SECURITY aims to publish thorough and cohesive overviews of specific topics in information security, as well as works that are larger in scope or that contain more detailed background information than can be accommodated in shorter survey articles. The series also serves as a forum for topics that may not have reached a level of maturity to warrant a comprehensive textbook treatment.

Researchers, as well as developers, are encouraged to contact Professor Sushil Jajodia with ideas for books under this series.

Additional titles in the series:

Additional information about this series can be obtained from
http://www.springeronline.com

PRIVACY PRESERVING DATA MINING

by

Jaideep Vaidya
Rutgers University, Newark, NJ

Chris Clifton
Purdue, W. Lafayette, IN, USA

Michael Zhu
Purdue, W. Lafayette, IN, USA

 Springer

Jaideep Vaidya
State Univ. New Jersey
Dept. Management Sciences &
    Information Systems
180 University Ave.
Newark NJ 07102-1803

Christopher W. Clifton
Purdue University
Dept. of Computer Science
250 N. University St.
West Lafayette IN 47907-2066

Yu Michael Zhu
Purdue University
Department of Statistics
Mathematical Sciences Bldg.1399
West Lafayette IN 47907-1399

Library of Congress Control Number: 2005934034

PRIVACY PRESERVING DATA MINING
by Jaideep Vaidya, Chris Clifton, Michael Zhu

ISBN-13: 978-0-387-25886-8
ISBN-10: 0-387-25886-7
e-ISBN-13: 978-0-387-29489-9
e-ISBN-10: 0-387-29489-6

Printed on acid-free paper.

Printed in the United States of America.

9 8 7 6 5 4 3 2 1   SPIN 11392194, 11570806

springeronline.com

To my parents and to Bhakti, with love.
−Jaideep

To my wife Patricia, with love.
−Chris

To my wife Ruomei, with love.
−Michael

Contents

Preface

Since its inception in 2000 with two conference papers titled "Privacy Preserving Data Mining", research on learning from data that we aren't allowed to see has multiplied dramatically. Publications have appeared in numerous venues, ranging from data mining to database to information security to cryptography. While there have been several privacy-preserving data mining workshops that bring together researchers from multiple communities, the research is still fragmented.

This book presents a sampling of work in the field. The primary target is the researcher or student who wishes to work in privacy-preserving data mining; the goal is to give a background on approaches along with details showing how to develop specific solutions within each approach. The book is organized much like a typical data mining text, with discussion of privacy-preserving solutions to particular data mining tasks. Readers with more general interests on the interaction between data mining and privacy will want to concentrate on Chapters 1-3 and 8, which describe privacy impacts of data mining and general approaches to privacy-preserving data mining. Those who have particular data mining problems to solve, but run into roadblocks because of privacy issues, may want to concentrate on the specific type of data mining task in Chapters 4-7.

The authors sincerely hope this book will be valuable in bringing order to this new and exciting research area; leading to advances that accomplish the apparently competing goals of extracting knowledge from data and protecting the privacy of the individuals the data is about.

West Lafayette, Indiana,   *Chris Clifton*

1

Privacy and Data Mining

Data mining has emerged as a significant technology for gaining knowledge from vast quantities of data. However, there has been growing concern that use of this technology is violating individual privacy. This has lead to a backlash against the technology. For example, a "Data-Mining Moratorium Act" introduced in the U.S. Senate that would have banned all data-mining programs (including research and development) by the U.S. Department of Defense[31]. While perhaps too extreme – as a hypothetical example, would data mining of equipment failure to improve maintenance schedules violate privacy? – the concern is real. There is growing concern over information privacy in general, with accompanying standards and legislation. This will be discussed in more detail in Chapter 2.

Data mining is perhaps unfairly demonized in this debate, a victim of misunderstanding of the technology. The goal of most data mining approaches is to develop generalized knowledge, rather than identify information about specific individuals. Market-basket association rules identify relationships among items purchases (e.g., "People who buy milk and eggs also buy butter"), the identity of the individuals who made such purposes are not a part of the result. Contrast with the "Data-Mining Reporting Act of 2003"[32], which defines data-mining as:

> (1) DATA-MINING- The term 'data-mining' means a query or search or other analysis of 1 or more electronic databases, where–
> (A) at least 1 of the databases was obtained from or remains under the control of a non-Federal entity, or the information was acquired initially by another department or agency of the Federal Government for purposes other than intelligence or law enforcement;
> (B) the search does not use a specific individual's personal identifiers to acquire information concerning that individual; and
> (C) a department or agency of the Federal Government is conducting the query or search or other analysis to find a pattern indicating terrorist or other criminal activity.

Note in particular clause (B), which talks specifically of searching for *information concerning that individual*. This is the opposite of most data mining, which is trying to move from information about individuals (the raw data) to generalizations that apply to broad classes. (A possible exception is Outlier Detection; techniques for outlier detection that limit the risk to privacy are discussed in Chapter 7.3.)

Does this mean that data mining (at least when used to develop generalized knowledge) does not pose a privacy risk? In practice, the answer is no. Perhaps the largest problem is not with data mining, but with the infrastructure used to support it. The more complete and accurate the data, the better the data mining results. The existence of complete, comprehensive, and accurate data sets raises privacy issues regardless of their intended use. The concern over, and eventual elimination of, the Total/Terrorism Information Awareness Program (the real target of the "Data-Mining Moratorium Act") was not because preventing terrorism was a bad idea – but because of the potential *misuse* of the data. While much of the data is already accessible, the fact that data is distributed among multiple databases, each under different authority, makes obtaining data for misuse difficult. The same problem arises with building data warehouses for data mining. Even though the data mining itself may be benign, gaining access to the data warehouse to misuse the data is much easier than gaining access to all of the original sources.

A second problem is with the results themselves. The census community has long recognized that publishing summaries of census data carries risks of violating privacy. Summary tables for a small census region may not identify an individual, but in combination (along with some knowledge about the individual, e.g., number of children and education level) it may be possible to isolate an individual and determine private information. There has been significant research showing how to release summary data without disclosing individual information[19]. Data mining results represent a new type of "summary data"; ensuring privacy means showing that the results (e.g., a set of association rules or a classification model) do not inherently disclose individual information.

The data mining and information security communities have recently begun addressing these issues. Numerous techniques have been developed that address the first problem – avoiding the potential for misuse posed by an integrated data warehouse. In short, techniques that allow mining when we aren't allowed to see the data. This work falls into two main categories: Data perturbation, and Secure Multiparty Computation. Data perturbation is based on the idea of not providing real data to the data miner – since the data isn't real, it shouldn't reveal private information. The data mining challenge is in how to obtain valid results from such data. The second category is based on separation of authority: Data is presumed to be controlled by different entities, and the goal is for those entities to cooperate to obtain valid data-mining results without disclosing their own data to others.

The second problem, the potential for data mining results to reveal private information, has received less attention. This is largely because concepts of privacy are not well-defined – without a formal definition, it is hard to say if privacy has been violated. We include a discussion of the work that has been done on this topic in Chapter 2.

Despite the fact that this field is new, and that privacy is not yet fully defined, there are many applications where privacy-preserving data mining can be shown to provide useful knowledge while meeting accepted standards for protecting privacy. As an example, consider mining of supermarket transaction data. Most supermarkets now offer discount cards to consumers who are willing to have their purchases tracked. Generating association rules from such data is a commonly used data mining example, leading to insight into buyer behavior that can be used to redesign store layouts, develop retailing promotions, etc.

This data can also be shared with suppliers, supporting their product development and marketing efforts. Unless substantial demographic information is removed, this could pose a privacy risk. Even if sufficient information is removed and the data cannot be traced back to the consumer, there is still a risk to the supermarket. Utilizing information from multiple retailers, a supplier may be able to develop promotions that favor one retailer over another, or that enhance supplier revenue at the expense of the retailer.

Instead, suppose that the retailers collaborate to produce globally valid association rules for the benefit of the supplier, without disclosing their own contribution to either the supplier or other retailers. This allows the supplier to improve product and marketing (benefiting all retailers), but does not provide the information needed to single out one retailer. Also notice that the individual data need not leave the retailer, solving the privacy problem raised by disclosing consumer data! In Chapter 6.2.1, we will see an algorithm that enables this scenario.

The goal of privacy-preserving data mining is to enable such win-win-win situations: The knowledge present in the data is extracted for use, the individual's privacy is protected, and the data holder is protected against misuse or disclosure of the data.

There are numerous drivers leading to increased demand for both data mining and privacy. On the data mining front, increased data collection is providing greater opportunities for data analysis. At the same time, an increasingly competitive world raises the cost of failing to utilize data. This can range from strategic business decisions (many view the decision as to the next plane by Airbus and Boeing to be make-or-break choices), to operational decisions (cost of overstocking or understocking items at a retailer), to intelligence discoveries (many believe that better data analysis could have prevented the September 11, 2001 terrorist attacks.)

At the same time, the costs of failing to protect privacy are increasing. For example, Toysmart.com gathered substantial customer information, promising that the private information would "never be shared with a third party."

When Toysmart.com filed for bankruptcy in 2000, the customer list was viewed as one of its more valuable assets. Toysmart.com was caught between the Bankruptcy court and creditors (who claimed rights to the list), and the Federal Trade Commission and TRUSTe (who claimed Toysmart.com was contractually prevented from disclosing the data). Walt Disney Corporation, the parent of Toysmart.com, eventually paid $50,000 to the creditors for the right to destroy the customer list.[64] More recently, in 2004 California passed SB1386, requiring a company to notify any California resident whose name and social security number, driver's license number, or financial information is disclosed through a breach of computerized data; such costs would almost certainly exceed the $.20/person that Disney paid to destroy Toysmart.com data.

Drivers for privacy-preserving data mining include:

- Legal requirements for protecting data. Perhaps the best known are the European Community's regulations [26] and the HIPAA healthcare regulations in the U.S. [40], but many jurisdictions are developing new and often more restrictive privacy laws.
- Liability from inadvertent disclosure of data. Even where legal protections do not prevent sharing of data, contractual obligations often require protection. A recent U.S. example of a credit card processor having 40 million credit card numbers stolen is a good example – the processor was not supposed to maintain data after processing was complete, but kept old data to analyze for fraud prevention (i.e., for data mining.)
- Proprietary information poses a tradeoff between the efficiency gains possible through sharing it with suppliers, and the risk of misuse of these trade secrets. Optimizing a supply chain is one example; companies face a tradeoff between greater efficiency in the supply chain, and revealing data to suppliers or customers that can compromise pricing and negotiating positions[7].
- Antitrust concerns restrict the ability of competitors to share information. How can competitors share information for allowed purposes (e.g., collaborative research on new technology), but still prove that the information shared does not enable collusion in pricing?

While the latter examples do not really appear to be a privacy issue, privacy-preserving data mining technology supports all of these needs. The goal of privacy-preserving data mining – analyzing data while limiting disclosure of that data – has numerous applications.

This book first looks more specifically at what is meant by privacy, as well as background in security and statistics on which most privacy-preserving data mining is built. A brief outline of the different classes of privacy-preserving data mining solutions, along with background theory behind those classes, is given in Chapter 3. Chapters 4–7 are organized by data mining task (classification, regression, associations, clustering), and present privacy-preserving data mining solutions for each of those tasks. The goal is not only to present algorithms to solve each of these problems, but to give an idea of the types of solutions that have been developed. This book does not attempt to present all the privacy-preserving data mining algorithms that have been developed. Instead, each algorithm presented introduces new approaches to preserving privacy; these differences are highlighted. Through understanding the spectrum of techniques and approaches that have been used for privacy-preserving data mining, the reader will have the understanding necessary to solve new privacy-preserving data mining problems.

2

What is Privacy?

A standard dictionary definition of privacy as it pertains to data is "freedom from unauthorized intrusion"[58]. With respect to privacy-preserving data mining, this does provide some insight. If users have given authorization to use the data for the particular data mining task, then there is no privacy issue. However, the second part is more difficult: If use is not authorized, what use constitutes "intrusion"?

A common standard among most privacy laws (e.g., European Community privacy guidelines[26] or the U.S. healthcare laws[40]) is that privacy only applies to "individually identifiable data". Combining *intrusion* and *individually identifiable* leads to a standard to judge privacy-preserving data mining: A privacy-preserving data mining technique must ensure that any information disclosed

1. cannot be traced to an individual; or
2. does not constitute an intrusion.

Formal definitions for both these items are an open challenge. At one extreme, we could assume that any data that does not give us completely accurate knowledge about a specific individual meets these criteria. At the other extreme, any improvement in our knowledge about an individual could be considered an intrusion. The latter is particularly likely to cause a problem for data mining, as the goal is to improve our knowledge. Even though the target is often groups of individuals, knowing more about a group does increase our knowledge about individuals in the group. This means we need to *measure* both the knowledge gained and our ability to relate it to a particular individual, and determine if these exceed thresholds.

This chapter first reviews metrics concerned with individual identifiability. This is not a complete review, but concentrates on work that has particular applicability to privacy-preserving data mining techniques. The second issue, what constitutes an intrusion, is less clearly defined. The end of the chapter will discuss some proposals for metrics to evaluate intrusiveness, but this is still very much an open problem.

To utilize this chapter in the concept of privacy-preserving data mining, it is important to remember that all disclosure from the data mining must be considered. This includes disclosure of data sets that have been altered/randomized to provide privacy, communications between parties participating in the mining process, and disclosure of the results of mining (e.g., a data mining model.) As this chapter introduces means of measuring privacy, examples will be provided of their relevance to the types of disclosures associated with privacy-preserving data mining.

2.1 Individual Identifiability

The U.S. Healthcare Information Portability and Accountability Act (HIPAA) defines *individually nonidentifiable data* as data "that does not identify an individual and with respect to which there is no reasonable basis to believe that the information can be used to identify an individual"[41]. The regulation requires an analysis that the risk of identification of individuals is very small in any data disclosed, *alone or in combination with other reasonably available information*. A real example of this is given in [79]: Medical data was disclosed with name and address removed. Linking with publicly available voter registration records using birth date, gender, and postal code revealed the name and address corresponding to the (presumed anonymous) medical records. This raises a key point: Just because the individual is not identifiable in the data is not sufficient; joining the data with other sources must not enable identification.

One proposed approach to prevent this is k-anonymity[76, 79]. The basic idea behind k-anonymity is to group individuals so that any identification is only to a group of k, not to an individual. This requires the introduction of a notion of *quasi-identifier*: information that can be used to link a record to an individual. With respect to the HIPAA definition, a quasi-identifier would be anything that would be present in "reasonably available information". The HIPAA regulations actually give a list of presumed quasi-identifiers; if these items are removed, data is considered not individually identifiable. The definition of k-anonymity states that any record must not be unique in its quasi-identifiers; there must be at least k records with the same quasi-identifier. This ensures that an attempt to identify an individual will result in at least k records that could apply to the individual. Assuming that the privacy-sensitive data (e.g., medical diagnoses) are not the same for all k records, then this throws uncertainty into any knowledge about an individual. The uncertainty lowers the risk that the knowledge constitutes an intrusion.

The idea that knowledge that applies to a group rather than a specific individual does not violate privacy has a long history. Census bureaus have used this approach as a means of protecting privacy. These agencies typically publish aggregate data in the form of contingency tables reflecting the count of individuals meeting a particular criterion (see Table 2.1). Note that some cells

Table 2.1. Excerpt from Table of Census Data, U.S. Census Bureau

Block Group 1, Census Tract 1, District
of Columbia, District of Columbia

Total:	9
Owner occupied:	3
1-person household	2
2-person household	1
. . .	
Renter occupied:	6
1-person household	3
2-person household	2
. . .	

list only a single such household. The disclosure problem is that combining this data with small cells in other tables (e.g., a table that reports salary by size of household, and a table reporting salary by racial characteristics) may reveal that only one possible salary is consistent with the numbers in all of the tables. For example, if we know that all owner-occupied 2-person households have salary over \$40,000, and of the nine multiracial households, only one has salary over \$40,000, we can determine that the single multiracial individual in an owner-occupied 2-person household makes over \$40,000. Since race and household size can often be observed, and home ownership status is publicly available (in the U.S.), this would result in disclosure of an individual salary.

Several methods are used to combat this. One is by introducing noise into the data; in Table 2.1 the Census Bureau warns that statistical procedures have been applied that introduce some uncertainty into data for small geographic areas with small population groups. Other techniques include cell suppression, in which counts smaller than a threshold are not reported at all; and generalization, where cells with small counts are merged (e.g., changing Table 2.1 so that it doesn't distinguish between owner-occupied and Renter-occupied housing.) Generalization and suppression are also used to achieve k-anonymity.

How does this apply to privacy-preserving data mining? If we can ensure that disclosures from the data mining generalize to large enough groups of individuals, then the size of the group can be used as a metric for privacy protection. This is of particular interest with respect to data mining results: When does the result itself violate privacy? The "size of group" standard may be easily met for some techniques; e.g., pruning approaches for decision trees may already generalize outcomes that apply to only small groups and association rule support counts provide a clear group size.

An unsolved problem for privacy-preserving data mining is the cumulative effect of multiple disclosures. While building a single model may meet the standard, multiple data mining models in combination may enable deducing individual information. This is closely related to the "multiple table" problem of census release, or the *statistical disclosure limitation* problem. Statistical disclosure limitation has been a topic of considerable study; readers interested in addressing the problem for data mining are urged to delve further into statistical disclosure limitation[18, 88, 86].

In addition to the "size of group" standard, the census community has developed techniques to measure risk of identifying an individual in a dataset. This has been used to evaluate the release of Public Use Microdata Sets: Data that appears to be actual census records for sets of individuals. Before release, several techniques are applied to the data: Generalization (e.g., limiting geographic detail), top/bottom coding (e.g., reporting a salary only as "greater than $100,000"), and data swapping (taking two records and swapping their values for one attribute.) These techniques introduce uncertainty into the data, thus limiting the confidence in attempts to identify an individual in the data. Combined with releasing only a sample of the dataset, it is likely that an identified individual is really a false match. This can happen if the individual is not in the sample, but swapping values between individuals in the sample creates a quasi-identifier that matches the target individual. Knowing that this is likely, an adversary trying to compromise privacy can have little confidence that the matching data really applies to the targeted individual.

A set of metrics are used to evaluate privacy preservation for public use microdata sets. One set is based on the value of the data, and includes preservation of univariate and covariate statistics on the data. The second deals with privacy, and is based on the percentage of individuals that a particularly well-equipped adversary could identify. Assumptions are that the adversary:

1. knows that some individuals are almost certainly in the sample (e.g., 600-1000 for a sample of 1500 individuals),
2. knows that the sample comes from a restricted set of individuals (e.g., 20,000),
3. has a good estimate (although some uncertainty) about the non-sensitive values (quasi-identifiers) for the target individuals, and
4. has a reasonable estimate of the sensitive values (e.g., within 10%.)

The metric is based on the number of individuals the adversary is able to correctly and confidently identify. In [60], identification rates of 13% are considered acceptably low. Note that this is an extremely well-informed adversary; in practice rates would be much lower.

While not a clean and simple metric like "size of group", this experimental approach that looks at the rate at which a well-informed adversary can identify individuals can be used to develop techniques to evaluate a variety of privacy-preserving data mining approaches. However, it is not amenable to a simple, "one size fits all" standard – as demonstrated in [60], applying this approach demands considerable understanding of the particular domain and the privacy risks associated with that domain.

There have been attempts to develop more formal definitions of anonymity that provide greater flexibility than k-anonymity. A metric presented in [15] uses the concept of anonymity, but specifically based on the ability to learn to distinguish individuals. The idea is that we should be unable to learn a classifier that distinguishes between individuals with high probability. The specific metric proposed was:

Definition 2.1. *[15] Two records that belong to different individuals I_1, I_2 are p-indistinguishable given data X if for every polynomial-time function $f : I \mapsto \{0, 1\}$*

$$|Pr\{f(I_1) = 1 | X\} - Pr\{f(I_2) = 1 | X\}| \leq p$$

where $0 < p < 1$.

Note the similarity to k-anonymity. This definition does not prevent us from learning sensitive information, it only poses a problem if that sensitive information is tied more closely to one individual rather than another. The difference is that this is a metric for the (sensitive) data X rather than the quasi-identifiers.

Further treatment along the same lines is given in [12], which defines a concept of isolation based on the ability of an adversary to "single out" an individual y in a set of points RDB using a query q:

Definition 2.2. *[12] Let y be any RDB point, and let $\delta_y = ||q - y||_2$. We say that q (c, t)-isolates y iff $B(q, c\delta_y)$ contains fewer than t points in the RDB, that is, $|B(q, c\delta_y) \cap RDB| < t$.*

The idea is that if y has at least t close neighbors, then anonymity (and privacy) is preserved. "Close" is determined by both a privacy threshold c, and how close the adversary's "guess" q is to the actual point y. With $c = 0$, or if the adversary knows the location of y, then k-anonymity is required to meet this standard. However, if an adversary has less information about y, the "anonymizing" neighbors need not be as close.

The paper continues with several sanitization algorithms that guarantee meeting the (c, t)-isolation standard. Perhaps most relevant to our discussion is that they show how to relate the definition to different "strength" adversaries. In particular, an adversary that generates a region that it believes y lies in versus an adversary that generates an action point q as the estimate. They show that there is essentially no difference in the ability of these adversaries to violate the (non)-isolation standard.

2.2 Measuring the Intrusiveness of Disclosure

To violate privacy, disclosed information must both be linked to an individual, and constitute an intrusion. While it is possible to develop broad definitions for individually identifiable, it is much harder to state what constitutes an intrusion. Release of some types of data, such as date of birth, pose only a minor annoyance by themselves. But in conjunction with other information date of birth can be used for identity theft, an unquestionable intrusion. Determining intrusiveness must be evaluated independently for each domain, making general approaches difficult.

What can be done is to measure the amount of information about a privacy sensitive attribute that is revealed to an adversary. As this is still an evolving area, we give only a brief description of several proposals rather than an in-depth treatment. It is our feeling that measuring intrusiveness of disclosure is still an open problem for privacy-preserving data mining; readers interested in addressing this problem are urged to consult the papers referenced in the following overview.

Bounded Knowledge.

Introducing uncertainty is a well established approach to protecting privacy. This leads to a metric based on the ability of an adversary to use the disclosed data to estimate a sensitive value. One such measure is given by [1]. They propose a measure based on the *differential entropy* of a random variable. The differential entropy $h(A)$ is a measure of the uncertainty inherent in A. Their metric for privacy is $2^{h(A)}$. Specifically, if we add noise from a random variable A, the privacy is:

$$\Pi(A) = 2^{-\int_{\Omega_A} f_A(a) log_2 f_A(a) da}$$

where Ω_A is the domain of A. There is a nice intuition behind this measure: The privacy is 0 if the exact value is known, and if the adversary knows only that the data is in a range of width a (but has no information on where in that range), $\Pi(A) = a$.

The problem with this metric is that an adversary may already have knowledge of the sensitive value; the real concern is how much that knowledge is increased by the data mining. This leads to a conditional privacy definition:

$$\Pi(A|B) = 2^{-\int_{\Omega_{A,B}} f_{A,B}(a,b) log_2 f_{A|B=b}(a) da db}$$

This was applied to noise addition to a dataset in [1]; this is discussed further in Chapter 4.2. However, the same metric can be applied to disclosures other than of the source data (although calculating the metric may be a challenge.)

A similar approach is taken in [14], where conditional entropy was used to evaluate disclosure from secure distributed protocols (see Chapter 3.3). While the definitions in Chapter 3.3 require perfect secrecy, the approach in [14] allows some disclosure. Assuming a uniform distribution of data, they are able to calculate the conditional entropy resulting from execution of a protocol (in particular, a set of linear equations that combine random noise and real data.) Using this, they analyze several scalar product protocols based on adding noise to a system of linear equations, then later factoring out the noise. The protocols result in sharing the "noisy" data; the technique of [14] enables evaluating the expected change in entropy resulting from the shared noisy data. While perhaps not directly applicable to all privacy-preserving data mining, the technique shows another way of calculating the information gained.

Need to know.

While not really a metric, the reason for disclosing information is important. Privacy laws generally include disclosure for certain permitted purposes, e.g. the European Union privacy guidelines specifically allow disclosure for government use or to carry out a transaction requested by the individual[26]:

> Member States shall provide that personal data may be processed only if:
> (a) the data subject has unambiguously given his consent; or
> (b) processing is necessary for the performance of a contract to which the data subject is party or in order to take steps at the request of the data subject prior to entering into a contract; or ...

This principle can be applied to data mining as well: disclose only the data actually needed to perform the desired task. We will show an example of this in Chapter 4.3. One approach produces a classifier, with the classification model being the outcome. Another provides the ability to classify, without actually revealing the model. If the goal is to classify new instances, the latter approach is less of a privacy threat. However, if the goal is to gain knowledge from understanding the model (e.g., understanding decision rules), then disclosure of that model may be acceptable.

Protected from disclosure.

Sometimes disclosure of certain data is specifically proscribed. We may find that *any* knowledge about that data is deemed too sensitive to reveal. For specific types of data mining, it may be possible to design techniques that limit ability to infer values from results, or even to control what results can be obtained. This is discussed further in Chapter 6.3. The problem in general is difficult. Data mining results inherently give knowledge. Combined with other knowledge available to an adversary, this may give *some* information about the protected data. A more detailed analysis of this type of disclosure will be discussed below.

Indirect disclosure.

Techniques to analyze a classifier to determine if it discloses sensitive data were explored in [48]. Their work made the assumption that the disclosure was a "black box" classifier – the adversary could classify instances, but not look inside the classifier. (Chapter 4.5 shows one way to do this.) A key insight of this work was to divide data into three classes: $\bar{S}$ensitive data, Public data, and data that is Unknown to the adversary. The basic metric used was the Bayes classification error rate. Assume we have data $(x_1, x_2, \ldots, x_n)$, that we want to classify x_i's into m classes $\{0, 1, \ldots, m-1\}$. For any classifier C:

$$x_i \mapsto C(x_i) \in \{0, 1, \ldots, m-1\}, \qquad i = 1, 2, \ldots, n,$$

we define the classifier accuracy for C as:

$$\sum_{i=0}^{m-1} Pr\{C(x) \neq i | z = i\} Pr\{z = i\}.$$

As an example, assume we have n samples $X = (x_1, x_2, \ldots, x_n)$ from a 2-point Gaussian mixture $(1 - \epsilon)N(0, 1) + \epsilon N(\mu, 1)$. We generate a sensitive data set $Z = (z_1, z_2, \ldots, z_n)$ where $z_i = 0$ if x_i is sampled from $N(0, 1)$, and $z_i = 1$ if x_i is sampled from $N(\mu, 1)$. For this simple classification problem, notice that out of the n samples, there are roughly ϵn samples from $N(\mu, 1)$, and $(1 - \epsilon)n$ from $N(0, 1)$. The total number of misclassified samples can be approximated by:

$$n(1 - \epsilon)Pr\{C(x) = 1 | z = 0\} + n\epsilon Pr\{C(x) = 0 | z = 1\};$$

dividing by n, we get the fraction of misclassified samples:

$$(1 - \epsilon)Pr\{C(x) = 1 | z = 0\} + \epsilon Pr\{C(x) = 0 | z = 1\};$$

and the metric gives the overall possibility that any sample is misclassified by C. Notice that this metric is an "overall" measure, not a measure for a particular value of x.

Based on this, several problems are analyzed in [48]. The obvious case is the example above: The classifier returns sensitive data. However, there are several more interesting cases. What if the classifier takes both public and unknown data as input? If we assume that all of the training data is known to the adversary (including public and sensitive, but not unknown, values), the classifier $C(P, U) \to S$ gives the adversary no additional knowledge about the sensitive values. But if the training data is unknown to the adversary, the classifier C does reveal sensitive data, *even though the adversary does not have complete information as input to the classifier.*

Another issue is the potential for privacy violation of a classifier that takes public data and discloses non-sensitive data to the adversary. While not in itself a privacy violation (no sensitive data is revealed), such a classifier could enable the adversary to deduce sensitive information. An experimental approach to evaluate this possibility is given in [48].

A final issue is raised by the fact that publicly available records already contain considerable information that many would consider private. If the private data revealed by a data mining process is already publicly available, does this pose a privacy risk? If the ease of access to that data is increased (e.g., available on the internet versus in person at a city hall), then the answer is yes. But if the data disclosed through data mining is as hard to obtain as the publicly available records, it isn't clear that the data mining poses a privacy threat.

Expanding on this argument, privacy risk really needs to be measured as the loss of privacy resulting from data mining. Suppose X is a sensitive attribute and its value for an fixed individual is equal to x. For example, $X = x$ is the salary of a professor at a university. Before any data processing and mining, some prior information may already exist regarding x. If each department publishes a range of salaries for each faculty rank, the prior information would be a bounded interval. Clearly, when addressing the impact of data mining on privacy, prior information also should be considered. Another type of external information comes from other attributes that are not privacy sensitive and are dependent on X. The values of these attributes, or even some properties regarding these attributes, are already public. Because of the dependence, information about X can be inferred from these attributes.

Several of the above techniques can be applied to these situations, in particular Bayesian inference, the conditional privacy definition of [1] (as well as a related conditional distribution definition from [27], and the indirect disclosure work of [48]. Still open is how to incorporate *ease of access* into these definitions.

3

Solution Approaches / Problems

In the current day and age, data collection is ubiquitous. Collating knowledge from this data is a valuable task. If the data is collected and mined at a single site, the data mining itself does not really pose an additional privacy risk; anyone with access to data at that site already has the specific individual information. While privacy laws may restrict use of such data for data mining (e.g., EC95/46 restricts how private data can be used), controlling such use is not really within the domain of privacy-preserving data mining technology. The technologies discussed in this book are instead concerned with preventing *disclosure* of private data: mining the data when we aren't allowed to see it. If individually identifiable data is not disclosed, the potential for intrusive misuse (and the resultant privacy breach) is eliminated.

The techniques presented in this book all start with an assumption that the source(s) and mining of the data are not all at the same site. This would seem to lead to distributed data mining techniques as a solution for privacy-preserving data mining. While we will see that such techniques serve as a basis for some privacy-preserving data mining algorithms, they do not solve the problem. Distributed data mining is effective when control of the data resides with a single party. From a privacy point of view, this is little different from data residing at a single site. If control/ownership of the data is centralized, the data could be centrally collected and classical data mining algorithms run. Distributed data mining approaches focus on increasing efficiency relative to such centralization of data. In order to save bandwidth or utilize the parallelism inherent in a distributed system, distributed data mining solutions often transfer summary information which in itself reveals significant information.

If data control or ownership is distributed, then disclosure of private information becomes an issue. This is the domain of privacy-preserving data mining. *How* control is distributed has a great impact on the appropriate solutions. For example, the first two privacy-preserving data mining papers both dealt with a situation where each party controlled information for a subset of individuals. In [56], the assumption was that two parties had the data divided between them: A "collaborating companies" model. The motivation for [4], individual survey data, lead to the opposite extreme: each of thousands of individuals controlled data on themselves. Because the way control or ownership of data is divided has such an impact on privacy-preserving data mining solutions, we now go into some detail on the way data can be divided and the resulting classes of solutions.

3.1 Data Partitioning Models

Before formulating solutions, it is necessary to first model the different ways in which data is distributed in the real world. There are two basic data partitioning / data distribution models: horizontal partitioning (a.k.a. homogeneous distribution) and vertical partitioning (a.k.a. heterogeneous distribution). We will now formally define these models. We define a dataset D in terms of the entities for whom the data is collected and the information that is collected for each entity. Thus, $D \equiv (E, I)$, where E is the entity set for whom information is collected and I is the feature set that is collected. We assume that there are k different sites, $P_1, \ldots, P_k$ collecting datasets $D_1 \equiv (E_1, I_1), \ldots, D_k \equiv (E_k, I_k)$ respectively.

Horizontal partitioning of data assumes that different sites collect the same sort of information about different entities. Therefore, in horizontal partitioning, $E_G = \bigcup_i E_i = E_1 \bigcup \ldots \bigcup E_k$, and $I_G = \bigcap_i = I_1 \bigcap \ldots \bigcap I_k$. Many such situations exist in real life. For example, all banks collect very similar information. However, the customer base for each bank tends to be quite different. Figure 3.1 demonstrates horizontal partitioning of data. The figure shows two banks, Citibank and JPMorgan Chase, each of which collects credit card information for their respective customers. Attributes such as the account balance, whether the account is new, active, delinquent are collected by both. Merging the two databases together should lead to more accurate predictive models used for activities like fraud detection.

On the other hand, vertical partitioning of data assumes that different sites collect different feature sets for the same set of entities. Thus, in vertical partitioning, $E_G = \bigcap_i E_i = E_1 \bigcap \ldots \bigcap E_k$, and $I_G = \bigcup_i = I_1 \bigcup \ldots \bigcup I_k$. For example, Ford collects information about vehicles manufactured. Firestone collects information about tires manufactured. Vehicles can be linked to tires. This linking information can be used to join the databases. The global database could then be mined to reveal useful information. Figure 3.2 demonstrates vertical partitioning of data. First, we see a hypothetical hospital / insurance company collecting medical records such as the type of brain tumor and diabetes (none if the person does not suffer from the condition). On the other hand, a wireless provider might be collecting other information such as the approximate amount of airtime used every day, the model of the cellphone and the kind of battery used. Together, merging this information for common customers and running data mining algorithms might give com-

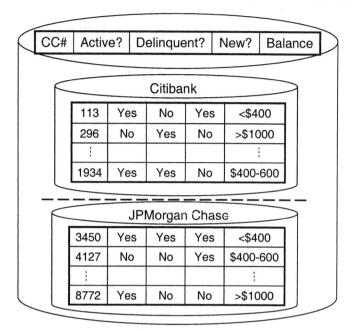

Fig. 3.1. Horizontal partitioning / Homogeneous distribution of data

pletely unexpected correlations (for example, a person with Type I diabetes using a cell phone with Li/Ion batteries for more than an hour per day is very likely to suffer from primary brain tumors.) It would be impossible to get such information by considering either database in isolation.

While there has been some work on more complex partitionings of data (e.g., [44] deals with data where the partitioning of each entity may be different), there is still considerable work to be done in this area.

3.2 Perturbation

One approach to privacy-preserving data mining is based on perturbating the original data, then providing the perturbed dataset as input to the data mining algorithm. The privacy-preserving properties are a result of the perturbation: Data values for individual entities are distorted, and thus individually identifiable (private) values are not revealed. An example would be a survey: A company wishes to mine data from a survey of private data values. While the respondents may be unwilling to provide those data values directly, they would be willing to provide perturbed/distorted results.

If an attribute is continuous, a simple perturbation method is to add noise generated from a specified probability distribution. Let X be an attribute and an individual have $X = x$, where x is a real value. Let r be a number

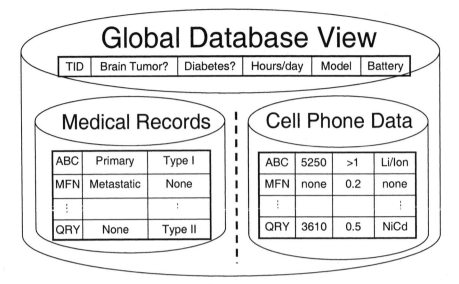

Fig. 3.2. Vertical partitioning / Heterogeneous distribution of data

randomly drawn from a normal distribution with mean 0 and variance 1. Instead of disclosing x, the individual reveals $x + r$. In fact, more complicated methods can be designed. For example, Warner [87] proposed the randomized response method for handling privacy sensitive questions in survey. Suppose an attribute Y with two values (yes or no) is of interest in a survey. The attribute however is private and an individual who participates the survey is not willing to disclose it. In stead of directly asking the question whether the surveyee has Y or not, the following two questions are presented:

1. I have the attribute Y.
2. I do not have the attribute Y.

The individual then use a randomizing device to decide which question to answer: The first is chosen with probability θ and the second question is chosen with probability $1 - \theta$. The surveyor gets either yes or no from the individual but does not know which question has been chosen and answered. Clearly, the value of Y thus obtained is the perturbed value and the true value or the privacy is protected. [23] used this technique for building privacy preserving decision trees. When mining association rules in market basket data, [28] proposed a a sophisticated scheme called the select-a-size randomization for preserving privacy, which will be discussed in detail in Section 6.1. Zhu and Liu [92] explored more sophisticated schemes for adding noise. Because randomization is usually an important part of most perturbation methods, we will use randomization and perturbation interchangeably in the book.

The randomized or noisy data preserves individual privacy, but it poses a challenge to data mining. Two crucial questions are how to mine the randomized data and how good the results based on randomized data are compared to the possible results from the original data. When data are sufficient, many aggregate properties can still be mined with enough accuracy, even when the randomization scheme is not exactly known. When the randomization scheme is known, then it is in generally possible to design a data mining tool in a way so that the best possible results can be obtained. It is understandable that some information or efficiency will be lost or compromised due to randomization. In most applications, the data mining tasks of interest are usually with a limited scope. Therefore, there is a possibility that randomization can be designed so that the information of interest can be preserved together with privacy, while irrelevant information is compromised. In general, the design of optimal randomization is still an open challenge.

Different data mining tasks and applications require different randomization schemes. The degree of randomization usually depends on how much privacy a data source wants to preserve, or how much information it allows others to learn. Kargupta et al. pointed out an important issue: arbitrary randomization is not safe [49]. Though randomized data may look quite different from the original data, an adversary may be able to take advantage of properties such as correlations and patterns in the original data to approximate their values accurately. For example, suppose a data contains one attribute and all its values are a constant. Based on the randomized data, an analyst can learn this fact fairly easily, which immediately results in a privacy breach. Similar situations will occur when the original data points demonstrate high sequential correlations or even deterministic patterns, or when the attributes are highly correlated. Huang el al. [42] further explore this issue as well and propose two data reconstruction methods based on data correlations – a Principal Component Analysis (PCA) technique and a Bayes Estimate (BE) technique. In general, data sources need to be aware of any special patterns in their data, and set up constraints that should be satisifed by any randomization schemes that they use. On the other hands, as discussed in the previous paragraph, excessive randomization will compromise the performance of a data mining algorithm or method. Thus, the efficacy of randomization critically depends on the way it is applied. For application, randomization schemes should be carefully designed to preserve a balance between privacy and information sharing and use.

3.3 Secure Multi-party Computation

Secure Multi-party Computation(SMC) refers to the general problem of secure computation of a function with distributed inputs. In general, any problem can be viewed as an SMC problem, and indeed all solution approaches fall under the broad umbrella of SMC. However, with respect to Privacy Preserving Data Mining, the general class of solutions that possess the rigor of work in SMC, and are typically based on cryptographic techniques are said to be

SMC solutions. Since a significant part of the book describes these solutions, we now provide a brief introduction to the field of SMC.

Yao first postulated the two-party comparison problem (Yao's Millionaire Protocol) and developed a provably secure solution[90]. This was extended to multiparty computations by Goldreich et al.[37]. They developed a framework for secure multiparty computation, and in [36] proved that computing a function privately is equivalent to computing it securely.

We start with the definitions for security in the semi-honest model. A semi-honest party (also referred to as honest but curious) follows the rules of the protocol using its correct input, but is free to later use what it sees during execution of the protocol to compromise security. A formal definition of private two-party computation in the semi-honest model is given below.

Definition 3.1. *(privacy with respect to semi-honest behavior):[36]*

Let $f : \{0,1\}^* \times \{0,1\}^* \longmapsto \{0,1\}^* \times \{0,1\}^*$ *be a functionality, and* $f_1(x,y)$ *(resp.,* $f_2(x,y)$*) denote the first (resp., second) element of* $f(x,y)$*. Let* Π *be two-party protocol for computing* f*. The* view *of the first (resp., second) party during an execution of* Π *on* (x,y)*, denoted* $\text{VIEW}_1^{\Pi}(x,y)$ *(resp.,* $\text{VIEW}_2^{\Pi}(x,y)$*), is* $(x,r,m_1,\ldots,m_t)$ *(resp.,* $(y,r,m_1,\ldots,m_t)$*), where* r *represents the outcome of the first (resp., second) party's internal coin tosses, and* m_i *represents the* i^{th} *message it has received. The* OUTPUT *of the first (resp., second) party during an execution of* Π *on* (x,y)*, denoted* $\text{OUTPUT}_1^{\Pi}(x,y)$ *(resp.,* $\text{OUTPUT}_2^{\Pi}(x,y)$*) is implicit in the party's own view of the execution, and* $\text{OUTPUT}^{\Pi}(x,y) = \left(\text{OUTPUT}_1^{\Pi}(x,y), \text{OUTPUT}_2^{\Pi}(x,y)\right)$*.*

(general case) We say that Π privately computes f *if there exist probabilistic polynomial-time algorithms, denoted* S_1 *and* S_2*, such that*

$$\{(S_1(x,f_1(x,y)), f(x,y))\}_{x,y} \stackrel{C}{\equiv} \left\{\left(\text{VIEW}_1^{\Pi}(x,y), \text{OUTPUT}^{\Pi}(x,y)\right)\right\}_{x,y}$$

$$\{(S_2(y,f_2(x,y)), f(x,y))\}_{x,y} \stackrel{C}{\equiv} \left\{\left(\text{VIEW}_2^{\Pi}(x,y), \text{OUTPUT}^{\Pi}(x,y)\right)\right\}_{x,y}$$

where $\stackrel{C}{\equiv}$ *denotes computational indistinguishability by (non-uniform) families of polynomial-size circuits.*

Privacy by Simulation

The above definition says that a computation is secure if the view of each party during the execution of the protocol can be effectively simulated given the input and the output of that party. Thus, in all of our proofs of security, we only need to show the existence of a simulator for each party that satisfies the above equations.

This does not quite guarantee that private information is protected. Whatever information can be deduced from the final result obviously cannot be kept private. For example, if a party learns that point A is an outlier, but point B which is close to A is not an outlier, it learns an estimate on the number of points that lie between the space covered by the hypersphere for A and hypersphere for B. Here, the result reveals information to the site having A and B. The key to the definition of privacy is that nothing is learned *beyond* what is inherent in the result.

A key result we use is the composition theorem. We state it for the semi-honest model. A detailed discussion of this theorem, as well as the proof, can be found in [36].

Theorem 3.2. *(Composition Theorem for the semi-honest model): Suppose that g is privately reducible to f and that there exists a protocol for privately computing f. Then there exists a protocol for privately computing g.*

Proof. Refer to [36].

The above definitions and theorems are relative to the semi-honest model. This model guarantees that parties who correctly follow the protocol do not have to fear seeing data they are not supposed to – this actually is sufficient for many practical applications of privacy-preserving data mining (e.g., where the concern is avoiding the cost of protecting private data.) The malicious model (guaranteeing that a malicious party cannot obtain private information from an honest one, among other things) adds considerable complexity. While many of the SMC-style protocols presented in this book do provide guarantees beyond that of the semi-honest model (such as guaranteeing that individual data items are not disclosed to a malicious party), few meet all the requirements of the malicious model. The definition above is sufficient for understanding this book; readers who wish to perform research in secure multiparty computation based privacy-preserving data mining protocols are urged to study [36].

Apart from the prior formulation, Goldreich also discusses an alternative formulation for privacy using the real vs. ideal model philosophy. A scheme is considered to be secure if whatever a feasible adversary can obtain in the real model, is also feasibly attainable in an ideal model. In this frame work, one first considers an ideal model in which the (two) parties are joined by a (third) trusted party, and the computation is performed via this trusted party. Next, one considers the real model in which a real (two-party) protocol is executed without any trusted third parties. A protocol in the real model is said to be secure with respect to certain adversarial behavior if the possible real executions with such an adversary can be "simulated" in the corresponding ideal model. The notion of simulation used here is different from the one used in Definition 3.1: Rather than simulating the view of a party via a traditional algorithm, the joint view of both parties needs to be simulated by the execution of an ideal-model protocol. Details can be found in [36].

3.3.1 Secure Circuit Evaluation

Perhaps the most important result to come out of the Secure Multiparty Computation community is a constructive proof that *any* polynomially computable function can be computed securely. This was accomplished by demonstrating that given a (polynomial size) boolean circuit with inputs split between parties, the circuit could be evaluated so that neither side would learn anything but the result. The idea is based on share splitting: the value for each "wire" in the circuit is split into two shares, such that the exclusive or of the two shares gives the true value. Say that the value on the wire should be 0 - this could be accomplished by both parties having 1, or both having 0. However, from one party's point of view, holding a 0 gives no information about the true value: we know that the other party's value is the true value, but we don't know what the other party's value is.

Andrew Yao showed that we could use cryptographic techniques to compute random shares of the output of a gate given random shares of the input, such that the exclusive or of the outputs gives the correct value. (This was formalized by Goldreich et al. in [37].) Two see this, let us view the case for a single gate, where each party holds one input. The two parties each choose a random bit, and provide the (randomly chosen) value r to the other party. They then replace their own input i with $i \oplus r$. Imagine the gate is an exclusive or: Party a then has $(i_a \oplus r_a)$ and r_b. Party a simply takes the exclusive or of these values to get $(i_a \oplus r_a) \oplus r_b$ as its share of the output. Party b likewise gets $(i_b \oplus r_b) \oplus r_a$ as its share. Note that neither has seen anything but a randomly chosen bit from the other party – clearly no information has been passed. However, the exclusive or of the two results is:

$$o = o_a \oplus o_b$$
$$= ((i_a \oplus r_a) \oplus r_b) \oplus ((i_b \oplus r_b) \oplus r_a)$$
$$= i_a \oplus i_b \oplus r_a \oplus r_a \oplus r_b \oplus r_b$$

Since any value exclusive orred with itself is 0, the random values cancel out and we are left with the correct result.

Inverting a value is also easy; one party simply inverts its random share; the other does nothing (both know the circuit, it is just the inputs that are private.) The and operation is more difficult, and involves a cryptographic protocol known as oblivious transfer. If we start with the random shares as described above, Party a randomly chooses its output o_a and constructs a table as follows:

$(i_b \oplus r_b), r_a$	0,0	0,1	1,0	1,1
o_b	$o_a + (i_a \oplus r_a) \cdot r_b$	$o_a + (i_a \oplus r_a) \cdot (r_b + 1)$	$o_a + ((i_a \oplus r_a) + 1) \cdot r_b$	$o_a + ((i_a \oplus r_a) + 1) \cdot (r_b + 1)$

Note that given party b's shares of the input (first line), the exclusive or of o_a with o_b (the second line) cancels out o_a, leaving the correct output for the gate. But the (randomly chosen) o_a hides this from Party b.

The cryptographic oblivious transfer protocol allows Party b to get the correct bit from the second row of this table, without being able to see any of the other bits or revealing to Party a which entry was chosen.

Repeating this process allows computing any arbitrarily large circuit (for details on the process, proof, and why it is limited to polynomial size see

[36].) The problem is that for data mining on large data sets, the number of inputs and size of the circuit become very large, and the computation cost becomes prohibitive. However, this method does enable efficient computation of functions of small inputs (such as comparing two numbers), and is used frequently as a subroutine in privacy-preserving data mining algorithms based on the secure multiparty computation model.

3.3.2 Secure Sum

We now go through a short example of secure computation to give a flavor of the overall idea – Secure sum. The secure sum problem is rather simple but extremely useful. Distributed data mining algorithms frequently calculate the sum of values from individual sites and thus use it as an underlying primitive.

The problem is defined as follows: Once again, we assume k parties, $P_1, \ldots, P_k$. Party P_i has a private value x_i. Together they want to compute the sum $S = \sum_{i=1}^{k} x_i$ in a secure fashion (i.e., without revealing anything except the final result). One other assumption is that the range of the sum is known (i.e., an upper bound on the sum). Thus, we assume that the sum S is a number in the field $\mathcal{F}$. Assuming at least 3 parties, the following protocol computes such a sum –

- P_1 generates a random number r from a uniform random distribution over the field $\mathcal{F}$.
- P_1 computes $S_1 = x_1 + r \bmod |F|$ and sends it to P_2
- For parties $P_2, \ldots, P_{k-1}$
    - P_i receives $S_{i-1} = r + \sum_{j=1}^{i-1} x_j \bmod |F|$.
    - P_i computes $S_i = S_{i-1} + x_i \bmod |F| = r + \sum_{j=1}^{i} x_j \bmod |F|$ and sends it to site P_{i+1}.
- P_k receives $S_{k-1} = r + \sum_{j=1}^{k-1} x_j \bmod |F|$.
- P_k computes $S_k = S_{k-1} + x_i \bmod |F| = r + \sum_{j=1}^{k} x_j \bmod |F|$ and sends it to site P_1.
- P_1 computes $S = S_k - r \bmod |F| = \sum_{j=1}^{k} x_j \bmod |F|$ and sends it to all other parties as well

Figure 3.3 depicts how this method operates on an example with 4 parties. The above protocol is secure in the SMC sense. The proof of security consists of showing how to simulate the messages received. Once those can be simulated in polynomial time, the messages sent can be easily computed. The basic idea is that every party (except P_1) only sees messages masked by a random number unknown to it, while P_1 only sees the final result. So, nothing new is learned by any party. Formally, P_i $(i = 2, \ldots, k)$ gets the message $S_{i-1} = r + \sum_{j=1}^{i-1} x_j$.

$$Pr(S_{i-1} = a) = Pr(r + \sum_{j=1}^{i-1} x_j = a) \qquad (3.1)$$

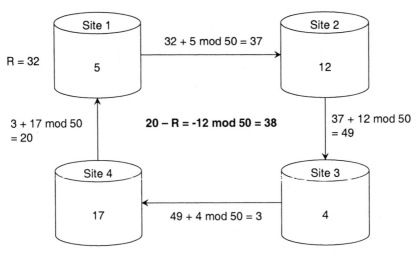

Fig. 3.3. Secure computation of a sum.

$$= Pr(r = a - \sum_{j=1}^{i-1} x_j) \tag{3.2}$$

$$= \frac{1}{|F|} \tag{3.3}$$

Thus, S_{i-1} can be simulated simply by randomly choosing a number from a uniform distribution over $\mathcal{F}$. Since P_1 knows the final result and the number r it chose, it can simulate the message it gets as well. Note that P_1 can also determine $\sum_{j=2}^{k} x_j$ by subtracting x_1. This is possible from the global result *regardless of how it is computed*, so P_1 has not learned anything from the computation.

In the protocol presented above, P_1 is designated as the initiator and the parties are ordered numerically (i.e., messages go from P_i to P_{i+1}. However, there is no special reason for either of these. Any party could be selected to initiate the protocol and receive the sum at the end. The order of the parties can also be scrambled (as long as every party does have the chance to add its private input).

This method faces an obvious problem if sites collude. Sites P_{l-1} and P_{l+1} can compare the values they send/receive to determine the exact value for x_l. The method can be extended to work for an honest majority. Each site divides x_l into shares. The sum for each share is computed individually. However, the path used is permuted for each share, such that no site has the same neighbor twice. To compute x_l, the neighbors of P_l from each iteration would have to collude. Varying the number of shares varies the number of dishonest (colluding) parties required to violate security.

One problem with both the randomization and cryptographic SMC approach is that unique secure solutions are required for every single data mining problem. While many of the building blocks used in these solutions are the same, this still remains a tremendous task, especially when considering the sheer number of different approaches possible. One possible way around this problem is to somehow transform the domain of the problem in a way that would make different data mining possible without requiring too much customization.

4

Predictive Modeling for Classification

Classification refers to the problem of categorizing observations into classes. Predictive modeling uses samples of data for which the class is known to generate a model for classifying new observations. Classification is ubiquitous in its applicability. Many real life problems reduce to classification. For example, medical diagnosis can be viewed as a classification problem: Symptoms and tests form the observation; the disease / diagnosis is the class. Similarly, fraud detection can be viewed as classification into fraudulent and non-fraudulent classes. Other examples abound.

There are several privacy issues associated with classification. The most obvious is with the samples used to generate, or learn, the classification model. The medical diagnosis example above would require samples of medical data; if individually identifiable this would be "protected healthcare information" under the U.S. HIPAA regulations. A second issue is with privacy of the observations themselves; imagine a "health self-checkup" web site, or a bank offering a service to predict the likelihood that a transaction is fraudulent. A third issue was discussed in Chapter 2.2: the classification model itself could be too effective, in effect revealing private information about individuals.

Example: Fraud Detection

To illustrate these issues, we will introduce an example based on credit card fraud detection. Credit card fraud is a burgeoning problem costing millions of dollars worldwide. Fair Isaac's Falcon Fraud Manager is used to monitor transactions for more than 450 million active accounts over six continents[30]. Consortium models incorporating data from hundreds of issuers have proven extremely useful in predicting fraud.

A key assumption of this approach is that Fair Isaac is trusted by all of the participating entities to keep their data secret from others. This imposes a high burden on Fair Isaac to ensure security of the data. In addition, privacy laws affect this model: many laws restrict trans-border disclosure of private information. (This includes transfer to the U.S., which has relatively weak privacy laws.)

A privacy-preserving solution would not require that actual private data be provided to Fair Isaac. This could involve ensemble approaches (card issuers provide a fraud model to Fair Isaac, rather than actual data), or having issues provide statistics that are not individually identifiable. Carrying this further, the card issuers may want to avoid having their own private data exposed. (Disclosure that an issuer had an unusually high percentage of fraudulent transactions would not be good for the stock price.) A full privacy-preserving solution would enable issuers to contribute to the development of the global fraud model, as well as use that model, without fear that their, or their customers', private data would be disclosed. Eliminating concerns over privacy could result in improved models: more sensitive data could be utilized, and entities that might otherwise have passed could participate.

Various techniques have evolved for classification. They include bayesian classification, decision tree based classification, neural network classification, and many others. For example, Fair Isaac uses an advanced neural network for fraud detection. In the most elemental sense, a classification algorithm trains a model out of the training data. In order to perform better than random, the algorithm computes some form of summary statistics from the training data, or encodes information in some way. Thus, inherently, some form of access to the data is assumed. Indeed most of the algorithms use the simplest possible means of computing these summary statistics through direct examination of data items. The privacy-preserving data mining problem, then, is to compute these statistics and construct the prediction model without having access to the data. Related to this is the issue of how the generated model is shared between the participating parties. Giving the global model to all parties may be appropriate in some cases, but not all. With a shared (privacy-preserving) model, some protocol is required to classify a new instance as well.

Privacy preserving solutions have been developed for several different techniques. Indeed, the entire field of privacy preserving data mining originated with two concurrently developed independent solutions for decision tree classification, emulating the ID3 algorithm when direct access to the data is not available.

This chapter contains a detailed view of privacy preserving solutions for ID3 classification, starting with a review of decision tree classification and the ID3 algorithm. We present three distinct solutions, each applicable to a different partitioning of the data. The two original papers in the field assumed horizontal partitioning, however one assumed that data was divided between two parties, while the other assumed that each individual provided their own data. This resulted in very different solutions, based on completely different models of privacy. Most privacy-preserving data mining work has build on one of the privacy models used in these original papers, so we will go into them in some detail. For completeness, we also introduce a solution for vertically partitioned data; this raises some new issues that do not occur with horizontal partitioning. We then discuss some of the privacy preserving solutions developed for other forms of classification.

4.1 Decision Tree Classification

Decision tree classification is one of the most widely used and practical methods for inductive inference. Decision tree learning is robust to noisy data and is capable of learning both conjunctive and disjunctive expressions. It is generally used to approximate discrete-valued target functions. Mitchell [59] characterizes problems suited to decision trees as follows (presentation courtesy Hamilton et al.[39]):

- Instances are composed of attribute-value pairs.
  - Instances are described by a fixed set of attributes (e.g., temperature) and their values (e.g., hot).
  - The easiest situation for decision tree learning occurs when each attribute takes on a small number of disjoint possible values (e.g., hot, mild, cold).
  - Extensions to the basic algorithm allow handling real-valued attributes as well (e.g., temperature).
- The target function has discrete output values.
  - A decision tree assigns a classification to each example. Boolean classification (with only two possible classes) is the simplest. Methods can easily be extended to learning functions multiple (> 2) possible output values.
  - Learning target functions with real-valued outputs is also possible (though significant extensions to the basic algorithm are necessary); these are commonly referred to as regression trees.
- Disjunctive descriptions may be required (since decision trees naturally represent disjunctive expressions).
- The training data may contain errors. Decision tree learning methods are robust to errors – both errors in classifications of the training examples and errors in the attribute values that describe these examples.
- The training data may contain missing attribute values. Decision tree methods can be used even when some training examples have unknown values (e.g., temperature is known for only some of the examples).

The model built by the algorithm is represented by a decision tree – hence the name. A decision tree is a sequential arrangement of tests (an appropriate test is prescribed at every step in an analysis). The leaves of the tree predict the class of the instance. Every path from the tree root to a leaf corresponds to a conjunction of attribute tests. Thus, the entire tree represents a disjunction of conjunctions of constraints on the attribute-values of instances. This tree can also be represented as a set of if-then rules. This adds to the readability and intuitiveness of the model.

For instance, consider the weather dataset shown in Table 4.1. Figure 4.1 shows one possible decision tree learned from this data set. New instances are classified by sorting them down the tree from the root node to some leaf node, which provides the classification of the instance. Every interior node of the tree specifies a test of some attribute for the instance; each branch descending from that node corresponds to one of the possible values for this attribute. So, an instance is classified by starting at the root node of the decision tree, testing the attribute specified by this node, then moving down the tree branch corresponding to the value of the attribute. This process is then repeated at the node on this branch and so on until a leaf node is reached. For example the instance $\{sunny, hot, normal, FALSE\}$ would be classified as "*Yes*" by the tree in figure 4.1.

Table 4.1. The Weather Dataset

outlook	temperature	humidity	windy	play
sunny	hot	high	FALSE	no
sunny	hot	high	TRUE	no
overcast	hot	high	FALSE	yes
rainy	mild	high	FALSE	yes
rainy	cool	normal	FALSE	yes
rainy	cool	normal	TRUE	no
overcast	cool	normal	TRUE	yes
sunny	mild	high	FALSE	no
sunny	cool	normal	FALSE	yes
rainy	mild	normal	FALSE	yes
sunny	mild	normal	TRUE	yes
overcast	mild	high	TRUE	yes
overcast	hot	normal	FALSE	yes
rainy	mild	high	TRUE	no

While many possible trees can be learned from the same set of training data, finding the optimal decision tree is an NP-complete problem. Occam's Razor (specialized to decision trees) is used as a guiding principle: "The world is inherently simple. Therefore the smallest decision tree that is consistent with the samples is the one that is most likely to identify unknown objects correctly". Rather than building all the possible trees, measuring the size of each, and choosing the smallest tree that best fits the data, several heuristics can be used in order to build a good tree.

Quinlan's ID3[72] algorithm is based on an information theoretic heuristic. It is appealingly simple and intuitive. As such, it is quite popular for constructing a decision tree. The seminal papers in Privacy Preserving Data Mining [4, 57] proposed solutions for constructing a decision tree using ID3 without disclosure of the data used to build the tree.

The basic ID3 algorithm is given in Algorithm 1. An information theoretic heuristic is used to decide the best attribute to split the tree. The subtrees are built by recursively applying the ID3 algorithm to the appropriate subset of the dataset. Building an ID3 decision tree is a recursive process, operating on

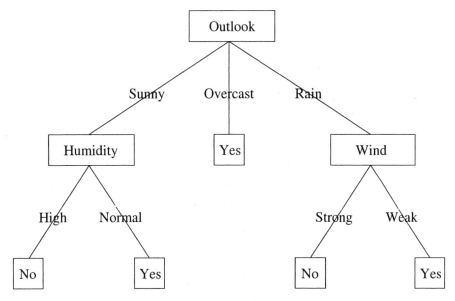

Fig. 4.1. A decision tree learned from the weather dataset

the decision attributes R, class attribute C, and training entities T. At each stage, one of three things can happen:

1. R might be empty; i.e., the algorithm has no attributes on which to make a choice. In this case, a decision on the class must be made simply on the basis of the transactions. A simple heuristic is to create a leaf node with the class of the leaf being the majority class of the transactions in T.
2. All the transactions in T may have the same class c. In this case, a leaf is created with class c.
3. Otherwise, we recurse:
   a) Find the attribute A that is the most effective classifier for transactions in T, specifically the attribute that gives the highest information gain.
   b) Partition T based on the values a_i of A.
   c) Return a tree with root labeled A and edges a_i, with the node at the end of edge a_i constructed from calling ID3 with $R - \{A\}, C, T(A_i)$.

In step 3a, *information gain* is defined as the change in the entropy relative to the class attribute. Specifically, the entropy

$$H_C(T) = \sum_{c \in C} -\frac{|T(c)|}{|T|} \log \frac{|T(c)|}{|T|}.$$

Analogously, the entropy after classifying with A is

$$H_C(T|A) = \sum_{a \in A} -\frac{|T(a)|}{|T|} H_C(T(a)).$$

Information gain due to the attribute A is now defined as

$$Gain(A) \stackrel{def}{=} H_C(T) - H_C(T|A).$$

The goal, then, is to find A that maximizes $Gain(A)$. Since $H_C(T)$ is fixed for any given T, this is equivalent to finding A that minimizes $H_C(T|A)$. Expanding, we get:

$$
\begin{aligned}
H_C(T|A) &= \sum_{a \in A} \frac{|T(a)|}{|T|} H_C(T(A)) \\
&= \frac{1}{|T|} \sum_{a \in A} |T(a)| \sum_{c \in C} -\frac{|T(a,c)|}{|T(a)|} \log\left(\frac{|T(a,c)|}{|T(A)|}\right) \\
&= \frac{1}{|T|} \left(-\sum_{a \in A} \sum_{c \in C} |T(a,c)| \log(|T(a,c)|) + \right. \\
&\qquad\qquad \left. \sum_{a \in A} |T(a)| \log(|T(a)|) \right)
\end{aligned}
\tag{4.1}
$$

Algorithm 1 ID3(R,C,T) tree learning algorithm

Require: R, the set of attributes
Require: C, the class attribute
Require: T, the set of transactions
1: **if** R is empty **then**
2:    return a leaf node, with class value assigned to most transactions in T
3: **else if** all transactions in T have the same class c **then**
4:    return a leaf node with the class c
5: **else**
6:    Determine the attribute A that best classifies the transactions in T
7:    Let $a_1, \ldots, a_m$ be the values of attribute A. Partition T into the m partitions $T(a_1), \ldots, T(a_m)$ such that every transaction in $T(a_i)$ has the attribute value a_i.
8:    Return a tree whose root is labeled A (this is the test attribute) and has m edges labeled $a_1, \ldots, a_m$ such that for every i, the edge a_i goes to the tree $ID3(R - A, C, T(a_i))$.
9: **end if**

4.2 A Perturbation-Based Solution for ID3

We now look at several perturbation based solutions for the classification problem. Recall that the focal processes of the perturbation based technique are

- the process of adding noise to the data
- the technique of learning the model from the noisy dataset

We start off by describing the solution proposed in the seminal paper by Agrawal and Srikant [4]. Agrawal and Srikant assume that the data is horizontally partitioned and the class is globally known. For example, a company wants a survey of the demographics of existing customers – each customer has his/her own information. Furthermore, the company already knows which are high-value customers, and wants to know what demographics correspond to high-value customers. The challenge is that customers do not want to reveal their demographic information. Instead, they give the company data that is perturbed by the addition of random noise. (As we shall see, while the added noise is random, it must come from a *distribution* that is known to the company.)

If we return to the description of ID3 in Section 4.1, we see that Steps 1 and 3c do not reference the (noisy) data. Step 2 references only the class data. Since this is assumed to be known, this only leaves Steps 3a and 3b: Finding the attribute with the maximum information gain and partitioning the tree based on that attribute. Looking at Equation 4.1, the only thing needed is $|T(a,c)|$ and $|T(a)|$.[1] $|T(a)|$ requires partitioning the entities based on the attribute value, exactly what is needed for Step 3b. The problem is that the attribute values are modified, so we don't know which entity really belongs in which partition.

Figure 4.2 demonstrates this problem graphically. There are clearly peaks in the number of drivers under 25 and in the 25-35 age range, but this doesn't hold in the noisy data. The ID3 partitioning should reflect the peaks in the data.

A second problem comes from the fact that the data is assumed to be ordered (otherwise "adding" noise makes no sense.) As a result, where to divide partitions is not obvious (as opposed to categorical data). Again, reconstructing the distribution can help. We can see that in Figure 4.2 partitioning the data at ages 30 and 50 would make sense – there is a natural "break" in the data at those points anyway. However, we can only see this from the actual distribution. The split points are not obvious in the noisy data.

Both these problems can be solved if we know the distribution of the original data, even if we do not know the original values. The problem remains that we may not get the *right* entities in each partition, but we are likely to get enough that the statistics on the class of each partition will still hold. (In [4] experimental results are given to verify this conjecture.)

What remains is the problem of estimating the distribution of the real data (X) given the noisy data (w) and the distribution of the noise (Y). This is accomplished through Bayes' rule:

---

[1] [4] actually uses the gini coefficient rather than information gain. While this may affect the quality of the decision tree, it has no impact on the discussion here. We stay with information gain for simplicity.

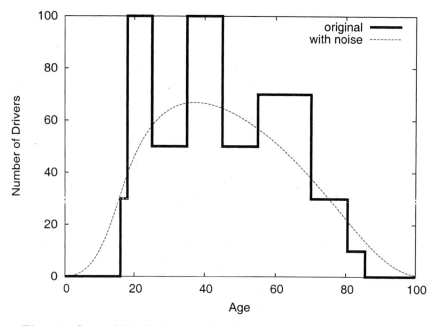

Fig. 4.2. Original distribution vs. distribution after random noise addition.

$$F_X'(a) \equiv \int_{-\infty}^{a} f_X(z|X + Y = w)dz$$

$$= \int_{-\infty}^{a} \frac{f_{X+Y}(w|X = z)f_X(z)}{f_{X+Y}(w)} dz$$

$$= \int_{-\infty}^{a} \frac{f_{X+Y}(w|X = z)f_X(z)}{\int_{-\infty}^{\infty} f_{X+Y}(w|X = z')f_X(z')dz'} dz$$

$$= \frac{\int_{-\infty}^{a} f_{X+Y}(w|X = z)f_X(z)dz}{\int_{-\infty}^{\infty} f_{X+Y}(w|X = z)f_X(z)dz}$$

$$= \frac{\int_{-\infty}^{a} f_Y(w - z)f_X(z)dz}{\int_{-\infty}^{\infty} f_Y(w - z)f_X(z)dz}$$

Given the actual data values $w_i = x_i + y_i$, we use this to estimate the distribution function as follows:

$$F_X'(a) = \frac{1}{n}\sum_{i=1}^{n} F_{X_i}' = \frac{1}{n}\sum_{i=1}^{n} \frac{\int_{-\infty}^{a} f_Y(w_i - z)f_X(z)dz}{\int_{-\infty}^{\infty} f_Y(w_i - z)f_X(z)dz}$$

Differentiating gives us the posterior density function:

$$f_X'(a) = \frac{1}{n}\sum_{i=1}^{n} \frac{f_Y(w_i - a)f_X(a)}{\int_{-\infty}^{\infty} f_Y(w_i - z)f_X(z)dz} \tag{4.2}$$

The only problem is, we don't know the real density function f_X. However, starting with an assumption of a uniform distribution, we can use Equation 4.2 to iteratively refine the density function estimate, converging on an estimate of the real distribution for X.

In [4] several optimizations are given, for example partitioning the data to convert the integration into sums. They also discuss tradeoffs in *when* to compute distributions: Once for each attribute? Separately for each class? For only the data that makes it to each split point? They found that reconstructing each attribute separately for each class gave the best performance/accuracy tradeoff, with classification accuracy substantially better than naïvely running on the noisy data, and approaching that of building a classifier directly on the real data.

One question with this approach is how much privacy is given? With the secure multiparty computation based approaches, the definition of privacy is clear. However, given a value that is based on the real value, how do we know how much noise is enough? Agrawal and Srikant proposed a metric based the confidence in estimating a value within a specified width: If it can be estimated with $c\%$ confidence that a value x lies in the interval $[x_l, x_h]$, then the privacy at the $c\%$ confidence level is $|x_h - x_l|$. They quantify this in terms of a percentage: The privacy metric for noise from a uniform distribution is the confidence times twice the interval width of the noise: 100% privacy corresponds to a 50% confidence that the values is within two distribution widths of the real value, or nearly 100% confidence that it is within one width. They have an equivalent definition for noise from a Gaussian distribution.

Agrawal and Aggarwal (not the same Agrawal) pointed out problems with this definition of privacy[1]. The very ability to reconstruct distributions may give us less privacy than expected. Figure 4.2 demonstrates this. Assume the noise is known to come from a uniform distribution over $[-15, 15]$, and the actual/reconstructed distribution is as shown by the bars. Since there are no drivers under age 16 (as determined from the reconstructed distribution), a driver whose age is given as 1 in the "privacy-preserving" dataset is known to be 16 years old – all privacy for this individual is lost. They instead give a definition based on entropy (discussed in Section 4.1). Specifically, if a random variable Y has entropy $H(Y)$, the privacy is $2^{H(Y)}$. This has the nice property that for a uniform distribution, the privacy is equivalent to the width of the interval from which the random value is chosen. This gives a meaningful way to compare different noise distributions.

They also provide a solution to the loss of privacy obtained through reconstructing the original data distribution. The idea is based on conditional entropy. Given the reconstructed distribution X, the privacy is now $2^{H(Y|X)}$. This naturally captures the expected privacy in terms of the interval width description: a reconstruction distribution that eliminates part of an interval (or makes it highly unlikely) gives a corresponding decrease in privacy.

4.3 A Cryptographic Solution for ID3

Lindell and Pinkas [56] were the first to propose privacy-preserving data mining using a cryptographic solution under the secure multiparty computation. They also targeted ID3 on horizontally partitioned data. Specifically, the solution in [57] assumes horizontal partitioning of data between two parties and shows how to build and ID3 tree. An interesting aspect of this solution is the ability to maintain "perfect" security in the SMC sense, while trading off efficiency against the quality of the resulting decision tree.

Revisiting the ID3 description from Section 4.1, we can assume that R and C are known to both parties. T is divided. In Step 1 we need only determine the class value of the majority of the transactions in T. This can be done using circuit evaluation (Chapter 3.3.1). Since each party is able to compute the count of local items in each class, the input size of the circuit is fixed by the number of classes, rather than growing with the (much larger) training data set size.

Step 2 requires only that we determine if all of the items are of the same class. This can again be done with circuit evaluation, here testing for equality. Each party gives as input either the single class c_i of all of its remaining items, or the special symbol $\perp$ if its items are of multiple classes. The circuit returns the input if the input values are equal, else it returns $\perp$.[2]

It is easy to prove that these two steps preserve privacy: Knowing the tree, we know the majority class for Step 1. As for Step 2, if we see a tree that has a "pruned" branch, we know that all items must be of the same class, or else the branch would have continued. Interestingly, if we test if all items are in the same class before testing if there are no more attributes (reversing steps 1 and 2, as the original ID3 algorithm was written), the algorithm would not be private. The problem is that Step 2 reveals if all of the items are of the same class. The decision tree doesn't contain this information. However, if a branch is "pruned" (the tree outputs the class without looking at all the attributes), we know that all the training data at that point are of the same class – otherwise the tree would have another split/level. Thus Step 2 doesn't reveal any knowledge that can't be inferred from the tree *when the tree is pruned* – the given order ensures that this step will only be taken if pruning is possible.

This leaves Step 3. Note that once A is known, steps 3b and 3c can be computed locally – no information exchange is required, so no privacy breach can occur. Since A can be determined by looking at the result tree, revealing A is not a problem, provided nothing but the proper choice for A is revealed. The hard part is Step 3a: computing the attribute that gives the highest information gain. This comes down to finding the A that minimizes Equation 4.1.

---

[2] The paper by Lindell and Pinkas gives other methods for computing this step, however circuit evaluation is sufficient – the readers are encouraged to read [57] for the details.

Note that since the database is horizontally partitioned, $|T(a)|$ is really $|T_1(a)| + |T_2(a)|$, where T_1 and T_2 are the two databases. The idea behind the privacy-preserving algorithm is that the parties will compute (random) shares of $(|T_1(a,c)| + |T_2(a,c)|) \log(|T_1(a,c)| + |T_2(a,c)|)$, and $(|T_1(a)| + |T_2(a)|) \log(|T_1(a)| + |T_2(a)|)$. The parties can then locally add their shares to give each a random share of $H_C(T|A)$. This is repeated for each attribute A, and a (small) circuit, of size linear in the number of attributes, is constructed to select the A that gives the largest value.

The problem, then is to efficiently compute $(x+y) \log(x+y)$. Lindell and Pinkas actually give a protocol for computing $(x+y) \ln(x+y)$, giving shares of $H_C(T|A) \cdot |T| \cdot \ln 2$. However, the constant factors are immaterial since the goal is simply to find the A that minimizes the equation. In [57] three protocols are given: Computing shares of $\ln(x+y)$, computing shares of $x \cdot y$, and the protocol for computing the final result. The last is straightforward: Given shares u_1 and u_2 of $\ln(x + y)$, the parties call the multiplication protocol twice to give shares of $u_1 \cdot y$ and $u_2 \cdot x$. Each party then sums three multiplications: the two secure multiplications, and the result of multiplying its input (x or y) with its share of the logarithm. This gives each shares of $u_1 y u_2 x + u_1 x + u_2 y = (x + y)(u_1 + u_2) = (x + y) \ln(x + y)$.

The logarithm and multiplication protocols are based on oblivious polynomial evaluation[62]. The idea of oblivious polynomial evaluation is that one party has a polynomial P, the other has a value for x, and the party holding x obtains $P(x)$ without learning P or revealing x. Given this, the multiplication protocol is simple: The first party chooses a random r and generates the polynomial $P(y) = xy - r$. The resulting of evaluating this on y is the second party's share: $xy - r$. The first party's share is simply r.

The challenge is computing shares of $\ln(x+y)$. The trick is to approximate $\ln(x + y)$ with a polynomial, specifically the Taylor series:

$$\ln(1 + \epsilon) = \sum_{i=1}^{k} \frac{(-1)^{i-1} \epsilon^i}{i}$$

Let 2^n be the closest power of 2 to $(x+y)$. Then $(x+y) = 2^n(1+\epsilon)$ for some $-1/2 \le \epsilon \le 1/2$. Now

$$\ln(x) = \ln(2^n(1 + \epsilon)) = n \ln 2 + \epsilon - \frac{\epsilon^2}{2} + \frac{\epsilon^3}{3} - \dots$$

We determine shares of $2^N n \ln 2$ and $2^N \epsilon$ (where N is an upper bound on n) using circuit evaluation. This is a simple circuit. $\epsilon \cdot 2^n = (x + y) - 2^n$, and n is obtained by inspecting the two most significant bits of $(x+y)$. There are a small (logarithmic in the database size) number of possibilities for $2^N n \ln 2$, and $\epsilon \cdot 2^N$ is obtained by left shifting $\epsilon \cdot 2^n$.

Assume the parties share of $2^N n \ln 2$ are α_1 and α_2, and the shares of $2^N \epsilon$ are β_1 and β_2. The first party defines

$$P(x) = \sum_{i=1}^{k} \frac{(-1)^{i-1}}{2^{N(i-1)}} \frac{(\alpha_1 + x)^i}{i} - r$$

and defines it's share $u_1 = \beta_1 + r$. The second party defines its share as $\beta_2 + P(\alpha_2)$. Note that $P(\alpha_2)$ computes the Taylor series approximation times 2^N, minus the random r. Since 2_N is public, it is easily divided out later, so the parties do get random shares of an approximation of $\ln(x + y)$.

As discussed in Section 3.3, all arithmetic is really done over a sufficiently large field, so that the random values (e.g., shares) can be chosen from a uniform distribution. In addition, the values in the Taylor series are multiplied by the least common multiple of $2, ..., k$ to eliminate fractions.

The key points to remember are the use of oblivious polynomial evaluation, and the use of an efficiently computable (bounded) approximation when efficiently and privately computing the real value is difficult.

4.4 ID3 on Vertically Partitioned Data

Quite different solutions are required for vertically partitioned data. A solution for constructing ID3 on vertically partitioned data was proposed by Du and Zhan[22]. Their work assumes that the data is vertically partitioned between two parties. The class of the training data is assumed to be shared, but some the attributes are private. Thus most steps can be evaluated locally. The main problem is computing which site has the best attribute to split on – each can compute the gain of their own attributes without reference to the other site. Instead of describing this in more detail, we explore a later solution proposed by Vaidya and Clifton that solves a more general problem – constructing an ID3 decision tree when the training data is vertically partitioned between many parties (≥ 2) and the class attribute is known to only a single party.

With vertically partitioned data, each party has knowledge of only some of the attributes. Thus, knowing the structure of the tree – especially, knowledge of an unknown attribute and its breakpoints for testing – constitutes a violation of the privacy of the individual parties.

In the best case, for no leakage of information, even the structure of the tree should be hidden, with an oblivious protocol for classifying a new instance. However, the cost associated with this is typically unacceptable. A compromise is to cloak the attribute tests used in the tree while still revealing the basic structure of the tree. A privacy-preserving decision tree is depicted in Figure 4.3. Note that each site only need know the branch values for the decisions it makes; e.g., site 1 would not know that the leftmost branch (performed by site 2) is based on humidity.

How do we go about constructing such a tree? Remember, that for vertically partitioned data, no single site knows R in its entirety. Instead each site i knows its own attributes R_i. A further assumption is that only one site knows the class attribute C. (Relaxing this assumption, as in [22], makes

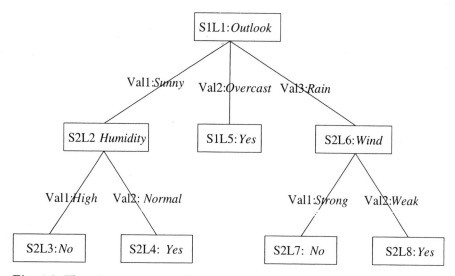

Fig. 4.3. The privacy preserving ID3 decision tree on the weather dataset (Mapping from identifiers to attributes and values is known only at the site holding attributes)

the problem simpler.) Every site also knows a *projection* of the transactions $\Pi_{R_i} T$. Each projection includes a transaction identifier that serves as a join key. (given in Algorithm 1). Privacy-preserving variants of the necessary steps are developed to ensure privacy of data.

Following the ID3 algorithm, the first challenge is to check if R is empty. This is based on Secure Sum (described in Chapter 3.3.2.) The key idea is to sum the number of remaining attributes at each site, and then compare with 0. If the sum is 0, clearly, no more attributes remain. While the secure summation protects the number of attributes available at any given site, the global sum is still revealed. To protect this,

- The first party adds a random r to its count of remaining items.
- This is passed to all sites, each adding its count.
- The last site and first then use commutative encryption to compare the final value to r (without revealing either) – if they are the same, R is empty.

Line 2 requires the determination of the majority class for a node. Since only one site knows the class, some protocol is required to do this. First, each site determines which of its transactions *might* reach that node of the tree. The intersection of these sets with the transactions in a particular class gives the number of transactions that reach that point in the tree having that particular class. Once this is done for all classes, the class site can now determine the distribution and majority class, and return a (leaf) node identifier. The identifier is used to map to the distribution at the time of classification.

The intersection process itself needs to be secure – this can be done by using a protocol for securely determining the cardinality of set intersection. Many protocols for doing so are known [84, 33, 2]. One of these protocols can be used.

To formalize the whole process, the notion of a *Constraint Set* is introduced. As the tree is being built, each party i keeps track of the values of its attributes used to reach that point in the tree in a filter $Constraints_i$. Initially, this is composed of all don't care values ('?'). However, when an attribute A_{ij} at site i is used (lines 6-7 of id3), entry j in $Constraints_i$ is set to the appropriate value before recursing to build the subtree. An example is given in Figure 4.4. The site has 6 attributes $A_1, \ldots, A_6$. The constraint tuple shows that the only transactions valid for this transaction are those with a value of 5 for A_1, *high* for A_2, and *warm* for A_5. The other attributes have a value of ? since they do not factor into the selection of an instance. Formally,

A_1	A_2	A_3	A_4	A_5	A_6
5	high	?	?	warm	?

Fig. 4.4. A constraint tuple for a single site

we define the following functions:

Constraints.set($attr, val$): Set the value of attribute $attr$ to val in the local constraints set. The special value '?' signifies a don't-care condition.

satisfies: x *satisfies* $Constraints_i$ if and only if the attribute values of the instance are compatible with the constraint tuple: $\forall i, (A_i(x) = v \Leftrightarrow Constraints(A_i) = v) \lor Constraints(A_i) = '?'$.

FormTransSet: *Function FormTransSet(Constraints): Return local transactions meeting constraints*

    $Y = \emptyset$
    for all transaction id $i \in T$ **do**
      if t_i satisfies $Constraints$ **then**
        $Y \leftarrow Y \cup \{i\}$
      end if
    end for
    return Y

Now, the majority class (and class distributions) are determined by computing for each class $\bigcap_{i=1..k} Y_i$, where Y_k includes a constraint on the class value.

DistributionCounts: *Function DistributionCounts(): Compute class distribution given current constraints*

- Every site except the class site forms the transaction set according to its local constraints
- for each class c_i
  - The class site includes a constraint for c_i in its local constraints set
  - The class site forms the transaction set according to the constraints
  - All sites execute a secure cardinality of set intersection protocol to find the count of transactions with the class c_i
- The class site returns the distribution of counts

The next issue is to determine if all transactions have the same class (Algorithm 1 line 3). If all are not the same class, as little information as possible should be disclosed. While this is possible without leaking any information, the associated cost in efficiency forces a compromise. For efficiency, the class site learns the count of classes even if this is an interior node; since it could compute this from the counts at the leaves of the subtree below the node, this discloses no additional information. The basic idea is simply to use the *DistributionCounts* function defined earlier to get the distribution and then check it. If all transactions are in the same class, we construct a leaf node. The class site maintains a mapping from the ID of that node to the resulting class distribution.

IsSameClass: *Function IsSameClass(): Find out if all transactions have the same class*
- Call the *DistributionCounts* function to get the distribution
- If only one of the counts is non-zero, build a leaf node with the distribution and return the ID of the node
- Otherwise return false

The next problem is to compute the best attribute: that with the maximum information gain. Once again, revisiting the equations from Section 4.1, the information gain when an attribute A is used to partition a data set T is:

$$Gain(A) = H_C(T) - H_C(T|A)$$
$$H_C(T) - \sum_{a \in A} \frac{|T(a)|}{|T|} H_C(T(A))$$

Remember, also that the entropy of a dataset T is given by:

$$H_C(T) = -\sum_{c \in C} \frac{|T(c)|}{|T|} \log \frac{|T(c)|}{|T|}$$

where $T(c)$ is the set of transactions having class c in T.

We can see that computing the gain for an attribute agains comes down to counting transactions. If the number of transactions reaching a node can be determined, the number in each class c, and the same two after partitioning with each possible attribute value $a \in A$, the gain due to A can be computed.

44 Predictive Modeling for Classification

Once this is done for all attributes, the attribute with the best information gain can be selected. The only tricky part of this is the fact that recursion cannot be used (at least directly), since no one knows the exact transaction set at a node. The constraint set is once again used to apply appropriate filters to get the correct count of transactions.

ComputeInfoGain: $ComputeInfoGain(A)$: *Compute the Information Gain for an attribute A*
- Get the total number of transactions at this node (using the DistributionCounts function)
- Compute the entropy of the transaction set at this node
- For each attribute value $a \in A$
  - Update the constraint set with the filter $A \to a$
  - Use the $DistributionCounts$ function to get the number of satisfying transactions
  - Compute the conditional entropy
- Compute the information gain

Finding the best attribute is a simple matter of finding out the information gain due to each attribute and selecting the best one. A naïve efficient implementation would leak the information gain due to each attribute. If even this minimal information should not be leaked, the information gain can be split between the parties, and a sequence of secure comparisons carried out to determine the best attribute.

AttribMaxInfoGain: $AttribMaxInfoGain()$: *return the site with the attribute having maximum information gain.*
- For each attribute A, compute the information gain using the function $ComputeInfoGain$.
- Find the attribute having the best information gain.

Once the best attribute has been determined, execution proceeds at that site. The site creates an interior node for the split, and then recurses.

The complete privacy-preserving distributed ID3 algorithm can be visualized as follows:

PPID3: $PPID3()$: *Privacy-preserving distributed ID3 over vertically partitioned data*
- If the attribute set is empty
  - Get the distribution counts
  - Create a leaf node and return the ID
- If all transactions have same class
  - Create a leaf node and return the ID
- Otherwise
  - Find the site with the attribute having the best information gain
  - At that site, create an interior node with the best attribute
  - For each value a for the best attribute A

    ·   create a local constraint $A \to a$

    ·   Recurse and add the appropriate branch to the inner node

– Unset the local constraint set for A

– Store the ID to node mapping locally and return the node ID

A model is of no use without a good way of applying it. The real goal of classification is to apply the model developed to predict the class of a new instance. With horizontally partitioned data, the developed model can be given to all sites, and any party can locally classify a new instance. With vertically partitioned data, the problem is more complex. Given that the structure of the tree is known (as in Figure 4.3), the root site first makes a decision based on its data. It then looks at the node this decision leads to and tells the site responsible for that node the node and the instance to be classified. This continues until a leaf is reached, and which point the site that originally held the class value knows the predicted class of the new instance. While this does lead to some disclosure of information (knowing the path followed, a site can say if instances have the same values for data not known to that site), specific values need not be disclosed.

4.5 Bayesian Methods

We now discuss the privacy preserving variants of several bayesian classification methods proposed in the literature. First up is Naïve Bayes Classification, followed by Bayesian Network structure learning.

Naïve Bayes is a simple but highly effective classifier. This combination of simplicity and effectiveness has lead to its use as a baseline standard by which other classifiers are measured. With various enhancements it is highly effective, and receives practical use in many applications (e.g., text classification[59]). Kantarcioglu and Vaidya[45] present a privacy-preserving solution for horizontally partitioned data while Vaidya and Clifton [82] do the same for vertically partitioned data. We now present a synopsis of both after briefly describing the Naïve Bayes classifier. This description is based on the discussion in Mitchell[59]. The Naïve Bayes classifier applies to learning tasks where each instance x is described by a conjunction of attribute values and the target function $f(x)$ can take on any value from some finite set C. A set of training examples of the target function is provided, and a new instance is presented, described by the tuple of attribute values $< a_1, a_2, \ldots, a_n >$. The learner is asked to predict the target value, or classification, for this new instance.

The Bayesian approach to classifying the new instance is to assign the most probable target value, c_{MAP}, given the attribute values $< a_1, a_2, \ldots, a_n >$ that describe the instance.

$$c_{MAP} = \underset{c_j \in C}{argmax} \left(P(c_j | a_1, a_2, \ldots, a_n) \right) \tag{4.3}$$

Using Bayes theorem,

$$c_{MAP} = \underset{c_j \in C}{argmax} \left(\frac{P(a_1, a_2, \ldots, a_n | c_j) P(c_j)}{P(a_1, a_2, \ldots, a_n)} \right)$$

$$= \underset{c_j \in C}{argmax} \left(P(a_1, a_2, \ldots, a_n | c_j) P(c_j) \right)$$

The Naïve Bayes classifier makes the further simplifying assumption that the attribute values are conditionally independent given the target value. Therefore,

$$c_{NB} = \underset{c_j \in C}{argmax} \left(P(c_j) \prod_i P(a_i | c_j) \right) \tag{4.4}$$

where c_{NB} denotes the target value output by the Naïve Bayes classifier.

The conditional probabilities $P(a_i | c_j)$ need to be estimated from the training set. The prior probabilities $P(c_j)$ also need to be fixed in some fashion (typically by simply counting the frequencies from the training set). The probabilities for differing hypotheses (classes) can also be computed by normalizing the values received for each hypothesis (class). Probabilities are computed differently for nominal and numeric attributes.

Nominal Attributes: For a nominal attribute X with r possible attributes values $x_1, \ldots, x_r$, the probability $P(X = x_k | c_j) = \frac{n_j}{n}$ where n is the total number of training examples for which $C = c_j$, and n_j is the number of those training examples that also have $X = x_k$.

Numeric Attributes: In the simplest case, numeric attributes are assumed to have a "normal" or "Gaussian" probability distribution.
The probability density function for a normal distribution with mean μ and variance σ^2 is given by

$$f(x) = \frac{1}{\sqrt{2\pi}\sigma} e^{-\frac{(x-\mu)^2}{2\sigma^2}} \tag{4.5}$$

The mean μ and variance σ^2 are calculated for each class and each numeric attribute from the training set. Now the required probability that the instance is of the class c_j, $P(X = x' | c_j)$, can be estimated by substituting $x = x'$ in equation 4.5.

In order to see how a privacy-preserving Naive Bayesian classifier is constructed, we need to address two issues: How to select the model parameters and how to classify a new instance.

The method in [82] is fully secure in the sense that even the model built is split between the participants. Thus, none of the participants knows the actual model parameters. The only information revealed is when a new instance is classified – the class of the instance. In contrast, for horizontally partitioned data, [45] propose that the global model built be shared by all participants so that each can locally classify a new instance as and when required. Let us now take a brief look at how this works.

4.5.1 Horizontally Partitioned Data

For horizontally partitioned data, it is possible to build very efficient protocols by compromising a little on security. In [45] an efficient protocol is achieved by allowing all parties to learn the total number of instances overall.

Building the model parameters

The procedure for building model parameters is different for nominal and numeric attributes.

Nominal attributes

For a nominal attribute, the conditional probability that an instance belongs to class c given that the instance has an attribute value $A = a$, $P(C = c|A = a)$, is given by

$$P(C = c|A = a) = \frac{P(C = c \cap A = a)}{P(A = a)} = \frac{n_{ac}}{n_a}.$$

n_{ac} is the number of instances in the (global) training set that have the class value c and an attribute value of a, while n_a is the (global) number of instances which simply have an attribute value of a. The necessary parameters are simply the counts of instances, n_{ac} and n_a.

With the data being horizontally partitioned, each party has partial information for every attribute. Each party can locally compute the counts for the instances local to it. The global count is simply the sum of these local counts. This global count can be easily computed securely using the secure sum protocol. Similarly, the global count of instances having the attribute value a can be computed through a secure sum of local counts. Thus, the required probability can be easily computed by dividing the appropriate global sums. The overall procedure is quite efficient since it only involves computing secure sums which can be done with negligible computation cost and linear communication cost.

Numeric attributes

For a numeric attribute, the necessary parameters are the mean μ and variance σ^2 for each class.

Again, the necessary information is split between all of the parties. To compute the mean, each party needs to sum the attribute values for all appropriate instances having the same class value. These local sums are added together and the global sum is divided by the total number of instances having that same class to get the mean for that class value. Thus, only the secure sum protocol is required to securely compute the means. Once all of the means μ_y are known, it is quite easy to compute the variance σ_y^2, for all class values.

Since each party knows the classification of the training instances it has, it can subtract the appropriate mean μ_y from an instance having class value y, square the value, and sum all such values together (again using secure sum). The global sum divided by the global number of instances having the same class y gives the required variance σ_y^2.

The vigilant (astute?) reader will notice that while efficient, the above protocols are not fully secure. Instead of just the mean and variance, more information is revealed. This additional information is actually the numerator and denominator in each case, instead of just the result of the division. As long as this information leakage is viewed acceptable, we have rather efficient protocols. However, if this leakage is deemed unacceptable, one must construct a more secure protocol. One way to do this is as follows – instead of completing the secure sums, the secure sum protocol leaves the sum split between two parties. These two parties can now engage in some sort of protocol for secure division to compute the final result. The interested reader can refer to [45] for more details.

Classifying a new instance

Since all the model parameters are known to all parties, evaluation is quite simple. The party thats needs to classify a new instance simply uses the Naïve Bayes evaluation procedure locally. Since the other parties have no participation in this, the question of privacy being compromised does not arise. An interesting point is that the model building algorithm can easily be modified to allow only one party to learn the model parameters. This party could then be the only one that could classify a new instance. This is useful in certain cases where one party simply wishes to use the training data present at several organizations to build a global classifier for itself.

4.5.2 Vertically Partitioned Data

Vertical partitioning of data has its own complications. A new issue is the location of the class attribute. There are two possibilities: the class may be known to all parties or it may be private to some party. This impacts the way the model is built and the way evaluation of a new instance is done. Both cases are realistic and model different situations. In the first case, each party can easily estimate all the required counts for nominal attributes and means and variances for numeric attributes locally, causing no privacy breaches. Prediction is also simple – each party can independently estimate the probabilities. All parties then securely multiply the probabilities and compare to obtain the predicted class. As such, we do not further discuss this. The other case is more challenging and is discussed in the following sections.

Building the model parameters

In this case, revealing the model parameters to everyone reveals too much information about the data of any one party. Since every attribute is private to some party – revealing its composition leaks too much information. One solution is to have the model parameters split between parties. An interactive protocol needs to be run to classify a new instance. The key is that the share of the parameters that each party gets appears to be random by itself. Only when added, do they have meaning. However, this addition occurs as part of the evaluation procedure which reveals only the class of a new instance. Let us now briefly look at the procedure for building model parameters for both nominal and numeric attributes.

Nominal attributes

Assume that party P_d holds the nominal attribute D, while party P_c holds the class attribute C. The necessary parameters are the probability estimates for every combination of attribute value and class value. However these need to be split between the two parties. Thus, if there are q classes and p attribute values, the goal is to compute $p \times q$ matrices S^c, S^d where the sum of corresponding entries $s_{li}^c + s_{li}^d$ gives the probability estimate for class c_i given that the attribute has value a_l.

A single probability estimate s_{li} can be computed as follows: P_d constructs a binary vector corresponding to the entities in the training set with 1 for each item having the value a_l and 0 for other items. P_c constructs a similar vector with $1/n_i$ for the n_i entities in the class, and 0 for other entities. The scalar product of the vectors gives the appropriate probability for the entry. Du and Atallah [20] show how to construct a secure protocol that return as output random shares of the scalar product. More detail on the entire procedure can be found in [82].

Numeric Attributes

For numeric attributes, computing the probability requires knowing the mean μ and variance σ^2 for each class value.

Computing the mean is similar to the earlier procedure for nominal attributes – for each class, P_c builds a vector of $1/n_i$ and 0 depending on whether the training entity is in the class or not. The mean for the class is the scalar product of this vector with the projection of the data onto the attribute. The scalar product protocol gives each party a share of the result, such that the sum is the mean (actually a constant times the mean, to convert to an integral value.)

Computing the variances is more difficult, as it requires summing the square of the distances between values and the mean, without revealing values to P_c or classes to P_d, or means to either. This is accomplished with homomorphic encryption: $E(a + b) = E(a) * E(b)$. P_d generates a homomorphic encryption key-pair. Next, P_d computes encrypted vectors of the data values and its share of the means and sends them to P_c, along with the encryption (but not decryption) key.

P_c takes each data value and subtracts the appropriate mean (both its share and the share sent by P_d) to give it the distance needed to compute the variance.

P_c also subtracts a random value, keeping the random value as its share of the distances. Homomorphic encryption makes this possible without decrypting. It now sends the vector back to P_d, which can decrypt to get the distance minus a random value. Thus, each has a random share of the distance.

The parties can now use a secure square computation protocol to compute shares of the square of the distance. Now, we are almost at the end. To find the variance for a particular class, we need to sum the square of the distances corresponding to instances having that class and then divide by the total number of instances having the class.

To do this, for each class value v_p, P_c creates the vector Y, where $Y_i = C/n_p$ if the ith transaction has class value v_p, otherwise $Y_i = 0$ (n_p is the total number of instances having class v_p). The scalar product of this vector with the share vectors computed earlier gives the shares of the variance for class value v_p (Actually, it gives shares of a constant times the variance, rather than the variance, but this can be adjusted for later in the evaluation procedure).

Classifying a new instance

Since all the model parameters are always split between some two parties, all of the parties must engage in a secure collaborative protocol to find the class of a new instance. This is quite complex and requires use of many subprotocols. We do not go into this in any depth other than stating that reasonably efficient solutions are possible. The interested reader is advised to look at [82] for more details.

4.5.3 Learning Bayesian Network Structure

The above methods support Naïve Bayes, which assumes that the impact of different attributes on the class is independent. Bayesian Networks relax this assumption, capturing where dependencies between attributes affect the class. A Bayesian Network is a graphical model; the vertices correspond to attributes, and the edges to probabilistic relationships between the attributes (Naï ve Bayes is thus a Bayesian Network with no edges.) The probability of a given class is similar to Equation 4.4, except that the probabilities associated with an attribute are conditional on the parents of that attribute in the network.

Wright and Yang [89] gave a privacy-preserving protocol for learning the Bayesian network structure for vertically partitioned data. This is a cryptographic protocol, as with the Naïve Bayes approach above, but specifically for two parties. Their approach is to emulate the $K2$ algorithm [16], which starts with a graph with no edges, then chooses a node and greedily adds the "parent" edge to that node that most improves a score for the network, stopping when a threshold for number of parents is reached.

Since the structure of the final network is presumed to be part of the outcome (and those not a privacy concern), the only issue is to determine which attribute most improves the score. Note the similarity with the decision tree protocol of Section 4.4; the difference is in the score function. Instead of information gain $K2$ algorithm uses:

$$f(i, \pi_i) = \prod_{j=1}^{q_i} \frac{d_i - 1)!}{\alpha_{ij} + d_i - 1)!} \prod_{k=1}^{d_i} \alpha_{ijk}! \qquad (4.6)$$

(For full details including the notation, please see [89]; our purpose here is not to give the full algorithm but to show the novel ideas with respect to privacy-preserving data mining.) The privacy-preserving solution works by first modifying the scoring function (taking the natural log of $f(i, \pi_i)$). While this changes the output, it doesn't affect the *order*; since all that matters is determining which attribute gives the highest score, the actual value is unimportant and the resulting network is unchanged. This same technique – transforming scoring functions in ways that do not alter the final result – has proven beneficial in designing other privacy-preserving data mining algorithms.

Note that by pushing the logarithm into Equation 4.6, the products turn into summations. Moreover, taking a page from [57] they approximate a difficult to compute value (in this case, Stirling's approximation for factorial.) Ignoring small factors in the approximation, the formula reduces to a sum of factors, where each factor is of the form $\ln x$ or $x \ln x$ (except for a final factor based on the number of possible values for each attribute, which they consider public knowledge.) This now reduces to secure summation and the $\ln x$ and $x \ln x$ protocols of [57].

4.6 Summary

We have seen several examples of how to create privacy-preserving classification algorithms. A common theme is to start with an algorithm and a partitioning of data. An appropriate model for preserving privacy is then chosen; If the data comes from many sources and noise is deemed sufficient to protect privacy, the data is perturbed. With fewer sources, cryptographic approaches are more appropriate. In either case, the algorithm is analyzed to determine what steps can be computed easily, and which are affected by privacy.

With data perturbation, the problem is generally determining distributions; the effect of noise on individual values will generally average out over time. With cryptographic protocols, the problem is generally computing a function from distributed inputs; through clever transformations to that function and use of a limited toolkit of techniques, it is often possible to efficiently evaluate or approximate the function without disclosing data. The number of algorithms that have been developed for privacy-preserving classification continues to grow. The methods presented here capture some of the key concepts that are being used in these algorithms.

5

Predictive Modeling for Regression

5.1 Introduction and Case Study

Regression is one of the most widely used statistical tools in data analysis and data mining. Suppose that a data set consists of a set of variables referred to as responses and another set of variables referred to as attributes (or independent variables). A central problem in data analysis is to model the relationship between the responses and the attributes. This is often termed as multiple responses multiple regression. A special case is multivariate regression – when the number of responses is one. In this chapter, we only deal with regression with one response. There are two primary goals of regression: one is to identify attributes that affect the response; the other is to predict the response for a new realization of the attributes, in other words, to fit a regression model.

There exist various types of regression that arise from different model assumptions. The simplest model assumption is to assume that the relationship between a response and a number of attributes is linear, which leads to linear regression. Linear regression is arguably the most popular and successful statistical method in data analysis. Nonlinear regression assumes an explicit nonlinear function, which depends on a number of unknown parameters, for the relationship between the response and the attributes. More generally, one does not want to or have to assume any specific form for the relationship. This leads to non-parametric regression. Many types of regression models exist between linear regression and general non-parametric regression, such as partial linear regression, additive model, etc. The choice of a proper regression model mostly depends on the application and the available data. While there have been umpteen different regression model looked at by statisticians, all of these have assumed that the basic data is freely available at a central site. In a privacy-preserving sense, the only work in this area has been on linear regression. Therefore, in this chapter, we focus only on linear regression.

Fitting a regression model is an art that consists of a series of steps including estimation, inferences, diagnostics, model selection, etc. The success of regression relies on the interaction between the analyst and data, and the role played by the analysts is much more crucial than one may realize. The complexity of regression also depends on the type of models one chooses to fit and the principle one uses to fit the model. The commonly used principles include least squares, maximum likelihood and non-parametric smoothing. Therefore, regression analysis is a dynamic and interactive procedure to build a model that relates a response and a number of attributes. Any competent regression software should be able to support all the activities required by this procedure. In this chapter, though we only consider the simplest model, linear regression, it still has most issues related to general regression. The development of secure linear regression would provide guidance for other more complicated types of regression in the future. For details regarding linear regression, readers are referred to Neter et al[63].

The necessity of developing privacy-preserving regression protocols arise when data are distributed among a number of data owners, or parties, who are interested in cooperatively identifying the relationship between a response and some attributes but are not willing to disclose their individual data to other owners due to concerns over data confidentiality and individual privacy. The data owners can be government agencies, corporations, etc. However, as every data owner only possesses part of the data, their private data may not be sufficient for each individual owner to fit the desired regression model properly. As before, we consider the following two modes of data separation:

Vertically partitioned data

Each data owner only holds a subset of the attributes and their values. In general, overlapping is allowed, which will not complicate the separation much. For convenience, we assume that the subsets held by the data owners are disjoint.

Horizontally partitioned data

Each data owner holds all of the attributes, however, he/she only owns a limited number of cases, which may not be enough for fitting an accurate regression model.

The responses might also be partitioned. For example, in horizontally partitioned data, every data owner holds the response for the cases in their individual data sets. In vertically partitioned data, just as in classification, the ownership of responses can be complicated. It is possible that only one owner, or a subset of owners, hold the responses, or the responses are shared by all the owners. To avoid unnecessary complexity, we assume that the data owners either hold the responses or are given the responses for vertically partitioned cases. We will briefly discuss the complications caused by the ownership of the responses for horizontally partitioned data.

Next, we will use a well-known real data set to illustrate the vertical and horizontal partitions of data in the regression setting.

5.1.1 Case Study

A well-known data set popularly used to illustrate linear regression is the Boston housing data first analyzed in Harrison and Rubinfel (1978). The original data contains 13 attributes that might affect housing values for 506 census tracts in the Boston standard statistical metropolitan area in 1970. The data is publicly available on the Internet. The attributes include Town (town name), Tract (tract ID number), Crim (per capita crime), Zn (proportions of residential land zoned for lots over 25000 sq.ft per town), Indus (proportion of non-retail business acres per town), Chas (binary with level 1 if tract borders Charles river; 0 otherwise), Nox (nitric oxides concentration per town), Rm (average numbers of rooms per dwelling), Age (proportion of owner-occupied units built prior to 1940), Dis (weighted distances to five Boston employment centers), Rad (an index of accessibility to radial highways), Tax (full-value property-tax rate per USD 10,000 per town), Ptration (pupil-teacher ration per town), B (transformed proportion of blacks), Lstat (percentage value of lower status population), and the response is Medv (median values of owner-occupied housing in USD 1000).

The data originated from several sources including the 1970 US census and the Boston Metropolitan Area Planning Committee. Hence, originally, the data was in fact distributed across several sources. For simplicity, we simply assume that there are only two sources (those mentioned above). If we further assume that the sources are not willing to share the data with each other, unlike 40 years ago, then the data should be considered as vertically partitioned. From a horizontal perspective, the data can considered as divided as tracts in Boston and tracts surrounding Boston. Data for these tracts could be held by different agents who are not willing to expose the data to other parties. In this case, we have horizontal partitioning of data. In general, a data can be partitioned both horizontally and vertically (in effect arbitrarily). In this chapter, however, the targeted data are either vertically partitioned or horizontally partitioned.

The purpose of regressing the Medv against other attributes is to identify the attributes that linearly affect the housing price in Boston. As expected, not all the attributes are important or significant. Hence, regression should be a dynamic and interactive procedure consisting of estimation or modeling fitting, diagnostics and model selection with the aim of deriving the best model.

5.1.2 What are the Problems?

For convenience, we assume that only two parties are engaged in regression. Following the convention in the literature, we name the first party Alice and the second party Bob. In the following, we deal with vertically partitioned data first and will discuss the horizontally partitioned case in a later section. Assume that Alice holds data for attributes $x_1, x_2, \ldots, x_n$ and Bob holds data for attributes $z_1, z_2, \ldots, z_m$, and both Alice and Bob know the response y. For simplicity, the attributes held by Alice and Bob do not overlap. Suppose N data points were observed. Let x_{ij} be the value of attribute x_j for the ith data point and similarly z_{ik} be the value of attribute z_k for the ith data point, where $1 \leq i \leq N$, $1 \leq j \leq m$ and $1 \leq k \leq n$. So Alice holds the data $(x_{ij})_{1 \leq i \leq N, 1 \leq j \leq n}$ and Bob the data $(z_{ik})_{1 \leq i \leq N, 1 \leq k \leq m}$. Suppose the response vector is $Y^\tau = (y_k)_{1 \leq k \leq N}$ that is known to both Alice and Bob, where τ means transpose. Alice and Bob desire to collectively identify and fit a proper regression model that approximates the relationship between y and some attributes, while not willing to or allowed to disclose their original data to each other. In the chapter, the focus is only on linear regression, and the attributes and their corresponding values, which form a column vector, are used interchangeably.

Let us consider a model involving attributes $x_{i_1}, x_{i_2}, \ldots, x_{i_s}$ (held by Alice) and attributes $z_{j_1}, z_{j_2}, \ldots, z_{j_t}$ (held by Bob). In general, higher order terms of these attributes such as $x_{i_j}^2$ are also candidate terms that need to be included in the model, so are the interactions between the attributes. The higher terms and the interactions between the attributes held by the same party can be regarded as induced new attributes and treated fairly easily. However, the interactions between the attributes held by different parties will pose challenges to privacy-preserving regression. For example, if we want to include the cross term $x_{i_1} z_{j_1}$ in the linear model, then it is necessary to generate a column vector equal to $x_{i_1} z_{j_1}$, which would result in immediate privacy breach if it is disclosed to both the parties. Certain extra measures must be exercised to prevent this from happening. Securely accommodating cross terms is a research issue that has not been treated or even addressed in the literature of privacy-preserving regression as of now. Hence, in the following discussion, we only discuss linear models that do not include cross terms. So, the linear model we consider is as follows,

$$y = \alpha_0 + \sum_{k=1}^{s} \alpha_{i_k} x_{i_k} + \sum_{l=1}^{t} \beta_{j_l} z_{j_l} + \epsilon \qquad (5.1)$$

where ϵ is a error term following a known distribution. For convenience, we assume that Alice always hold the N-dimensional column vector whose entries are identical to 1, which is denoted by x_{i_0}. Then the model becomes

$$y = \sum_{k=0}^{s} \alpha_{i_k} x_{i_k} + \sum_{l=1}^{t} \beta_{j_l} z_{j_l} + \epsilon \qquad (5.2)$$

where we assume that $\alpha_{i_0} = \alpha_0$. Let $A = (x_{i_0}, x_{i_1}, \ldots, x_{i_s})$ be a $N \times (s+1)$ matrix containing the values of $x_{i_1}, \ldots, x_{i_s}$, and $B = (z_{j_1}, \ldots, z_{j_t})$ a $N \times t$ matrix containing the values of $z_{j_1}, \ldots, z_{j_t}$. Hence, the data matrix for the model above is $X = (A, B)$. Let $\beta_A^\tau = (\alpha_0, \alpha_{i_1}, \ldots, \alpha_{i_s})$ and $\beta_B^\tau = (\beta_{j_1}, \ldots, \beta_{j_t})$. Then $\beta^\tau = (\beta_A^\tau, \beta_B^\tau)$ is the vector of regression coefficients. Recall that Y is the vector of responses. It is well-known that the least squares estimate of β is

$$\hat{\beta} = (X^{\tau}X)^{-1}X^{\tau}Y. \qquad (5.3)$$

Although the calculation of $\hat{\beta}$ is an important step in regression, it is not the only step. In fact it is a relatively easy first step. Other steps including diagnostics and model selection are even more important and challenging. Diagnostics are used to assess whether a fixed model is proper and suggest further analyses towards the best regression model and results. Usually model diagnostics rely on either statistics measuring the goodness of fit of the obtained model or graphical tools. Diagnostics are also used to discover peculiarities such as outliers that exists in the data. Statistics reflecting the goodness of fit of a model include correlation coefficient, R^2, adjusted R^2 (or R_a^2), etc. Graphical tools include the residual versus predicted response plot and the residual versus predictor plots. These plots may suggest that certain model assumptions are violated and need to be remedied using transformation or suggest that cross terms and high order terms of the involved attributes should be added to the model. Clearly, residuals play an essential role in diagnostics. Once $\hat{\beta}$ is available, we can calculate the fitted or predicted responses as follows,

$$\hat{Y} = (\hat{y}_1, \ldots, \hat{y}_N)^{\tau} = X\hat{\beta}. \qquad (5.4)$$

The column vector of residues is $\hat{\varepsilon} = Y - \hat{Y}$ and the residual for the ith data point is $\hat{\varepsilon}_i = y_i - \hat{y}_i$. Then

$$R^2 = 1 - \frac{\sum_{i=1}^{N}(y_i - \hat{y}_i)^2}{\sum_{i=1}^{N}(y_i - \bar{y})^2} \qquad (5.5)$$

and

$$R_a^2 = 1 - \frac{\sum_{i=1}^{N}(y_i - \hat{y}_i)^2/(N - p - 1)}{\sum_{i=1}^{N}(y_i - \bar{y})^2/(N - 1)} \qquad (5.6)$$

where $p = s + t$ is the number of attributes involved in the model. Using the residuals, the residual plots can be easily generated. For example, the residual versus predicted response plot is the plot of $\hat{\varepsilon}_i$ against $\hat{y}_i$ for $1 \leq i \leq N$. Diagnostic measures and plots may also reveal the original data, thus posing a challenge to privacy preserving regression. As will be discussed later on, these issues have not been well addressed and treated in the literature.

The other step we want to discuss is model selection in regression. Note that in the linear model above, only s of the m attributes held by Alice and t of the n attributes held by Bob are involved. As discussed earlier, the ultimate goal of regression is to identify a subset of the $m + n$ attributes and the best regression model including these attributes. This procedure is referred to as model selection and should be regarded as the most important step in regression. The key of model selection is to discriminate different regression models involving different sets of attributes. Some diagnostic statistics such as those measuring the goodness of fit can be used for model selection, but much caution should be exerted. For example, the full model that includes all the attributes attains the maximum R^2 among the models without high order and cross terms, but the full model is very likely a poor model because many irrelevant attributes are included and it overfits the data. The fitted full model usually performs poorly in terms of prediction.

In statistics, various statistics or criteria are available to distinguish two different models. For two nested models, the extra sum of squares principle can be used to decide whether including extra attributes or terms in the smaller model is statistically necessary. In the following, we would like to briefly introduce one of the most popularly used model selection criteria called Mallow's C_p. Let RSS_{full} be the sum of squared residuals generated from the regression with the full model, and let $RSS_{s,i}$ be the sum of squared residuals generated from the regression with the chosen attributes in (5.?). Then

$$C_p = \frac{RSS_{s,t}}{\hat{\sigma}^2} - (n - 2p), \tag{5.7}$$

where

$$\hat{\sigma}^2 = \frac{RSS_{full}}{N - m - n - 1}.$$

Varying the chosen attributes, different regression models can be fitted and their corresponding C_p can be calculated. In the end, the models that correspond to a C_p approximately equal to p are considered to be the best models. Other model selection procedures include step-wise regression and best subset regression. It should be clear from the above discussion that process of regression is not simply one step estimation of the regression coefficients. In fact, many models, sometimes in the order of hundreds or thousands, need to be fitted and diagnosed to arrive at the best model. Facilitating this complex procedure while preserving privacy in much more challenging than the secure estimation of the regression coefficients of a fixed model.

Many other issues are also present in regression such as outlier detection, high collinearity, insufficient-rank etc. To resolve these issues, regularization and other robust techniques are needed. They may also cause security concerns for privacy preserving regression. For now, however, in this chapter we will only focus on a much simplified regression procedure that consists of estimation, diagnostics and model selection. Most results in the privacy-preserving literature touch only the first step, but we stress that the other two steps deserve attention and further research in the future. It is our belief that only when the issues of privacy preservation in these three basic steps have been tackled well, can a *practically* useful privacy-preserving regression protocol be built and tested.

5.1.3 Weak Secure Model

As the discussion in Section 3.3 shows, general multiparty computation is theoretically feasible. The theoretical work of Goldreich et al. is based on circuit evaluation over finite field, specifically involving boolean computations. Although general regression is a complicated procedure, in theory it can always be decomposed into a collection of basic or atomic operations of Boolean nature. Hence, theoretically, we can conclude that secure multiparty computation is feasible for regression. However, although the statement is appealing, it does not convey any practical means and relief. Naïve application of the circuit evaluation technique makes the protocol way too complex to be useful in practice. In addition, even though every atomic operation is secure, the security of the whole procedure may not be guaranteed (if intermediate results are revealed). More practical solutions are clearly required.

The definitional approach to SMC is very appealing in terms of security since it clearly expresses what it means for a protocol to be secure. However, this often is too rigorous and it is difficult to formulate solutions for practical problems that are efficient *and* meet with the security definitions. There has been an effort to propose weaker models which would allow more efficient solutions to be proposed. But this is still immature and in progress.

Du et al. [21] propose a weak security model under which certain protocols might hold. Although it is still very heuristic and primitive, it does serve as a starting point which could be further improved. We now present the model. The model is defined for two parties Alice and Bob as follows:

Weak Security Model

Let I_A and I_B be the private inputs of A and B respectively, and O_A and O_B be their respective outputs for the function $\mathcal{F}$; i.e., $(O_A, O_B) = \mathcal{F}(I_A, I_B)$. A protocol C for computing $\mathcal{F}$ is secure against dishonest Bob if there exist an infinite number of (I'_A, O'_A) pairs in $(\mathcal{R}, \mathcal{R})$ such that $(O'_A, O_B) = \mathcal{F}(I'_A, I_B)$. Similarly, C is secure against dishonest Alice if there exist an infinite number of pairs (I'_B, O'_B) such that $(O_A, O'_B) = C(I_A, I'_B)$.

Intuitively, under this security model a protocol is considered to be secure, if for any input/output pair (I, O) from one party, there exist an infinite number of possible input/output pairs in $\mathcal{R}$ for the other party such that the result of the protocol is still O for the first party when its input is I. Therefore, from its own observed output, a party cannot determine the inputs from the other party.

The model above is rather primitive and much improvement is necessary. For example, as Du et al point out, even if a protocol C is secure against Bob under the model above, if the infinite number of input/output pairs of Alice are all concentrated around the true input and output, then Bob can get clear knowledge of the information of Alice, thus incurring a privacy breach. One possibility to improve the model might be to incorporate a distribution or a distance measure between all the possible pairs.

However, other more serious problems still exist. The model is not stated rigorously enough. For example, the model does not state anything about what is revealed during the protocol. So, a protocol C might simply reveal A's input

I_A during computation. While there may be infinite possible (I'_A, O'_A) pairs possible, B would know A's actual input, while the protocol would be considered safe under the proposed security model. In our opinion, this security model has much better applicability for examining the question of whether the results reveal information about the input and could be used in conjunction with the standard SMC model. In any case, keeping these problems in mind, the model serves as a heuristic start point, but clearly needs further scrutiny and analysis.

5.2 Vertically Partitioned Data

5.2.1 Secure Estimation of Regression Coefficients

Recall that, in Section 5.1.2, A is the data matrix for the attributes held by Alice and B the data matrix for the attributes held by Bob. In terms of A and B, the formula for calculating the estimate $\hat{\beta}$ is

$$\begin{pmatrix} \hat{\alpha}_A \\ \hat{\beta}_B \end{pmatrix} = \begin{pmatrix} A^\tau A & A^\tau B \\ B^\tau A & B^\tau B \end{pmatrix}^{-1} \begin{pmatrix} A^\tau Y \\ B^\tau Y \end{pmatrix} \tag{5.8}$$

It is clear that matrix product plays a central role in calculating the regression coefficients, and security is of concern when matrices from different parties are involved. For example, we need a secure way for calculating $A^\tau B$. Indeed, it is easy to see that if secure protocols for matrix multiplication and matrix inversion are available, it is quite easy to create a secure protocol for estimating the regression coefficients.

In order to facilitate these calculations, some basic protocols were developed in the literature. In the following, we introduce some of these protocols first, then show how they can be employed to securely calculate the coefficients.

The first protocol was proposed by Du et al[21]. This protocol assumes the use of a Commodity Server, a third party that helps the two parties in computing their goal. The commodity server is considered to be semi-trusted, and it does not collude with either of the parties. The main advantage of utilizing a commodity server is to make the secure computation significantly more efficient. The commodity server does not learn any information from the protocol, and it only provides necessary help to make the secure computation possible. The secure matrix product with a CS is described as follows:

Secure $A^\tau \cdot B$ protocol with commodity server

1. The commodity Sever generates random matrices R_a, R_b and r_a of dimensions $N \times m$, $N \times n$ and $m \times n$, respectively; let $r_b = R_a^\tau R_b - r_a$. The server then sends (R_a, r_a) to Alice and (R_b, r_b) to Bob;
2. Alice sends $\tilde{A} = A + R_a$ to Bob, and Bob sends $\tilde{B} = B + R_b$ to Alice;

3. Bob generates a $m \times n$ random matrix V_b and compute $T = \tilde{A} \cdot B + (r_b - V_b)$, then send it to Alice;
4. Alice computes $V_a = T + r_a - R_a^\tau \cdot B$.

It can be easily verified that $V_a + V_b = A^\tau \cdot B$ and [21] proved that neither party can learn about the original data of the other party under their model. Therefore, the product of two matrices is calculated without exposing the raw data. Note, however, that it is rather difficult to prove the security of the model under the standard SMC definitions of security. Clearly, the success of this protocol relies on the participating parties being semi-honest. Since the commodity server only provides necessary random matrices, it cannot learn anything about A and B unless it colludes with Alice or Bob. In practice, it may be difficult to find a trustworthy commodity server who can be relied upon not to collude with one of the parties. In this case, one should use only a 2-party secure protocol. [21] also proposed the following solution for secure matrix product. For convenience, we assume that N is even. Suppose M is an arbitrary $N \times N$ matrix with entries m_{ij} and $M^{-1} = (m^{ij})$ is the inverse of M. Let $M_{\text{left}} = (m_{ij})_{1 \leq i \leq N, 1 \leq j \leq N/2}$, $M_{\text{right}} = (m_{ij})_{1 \leq i \leq N, N/2+1 \leq j \leq N}$, $M_{\text{top}}^{-1} = (m^{ij})_{1 \leq i \leq N/2, 1 \leq j \leq N}$, $M_{\text{bottom}}^{-1} = (m^{ij})_{N/2+1 \leq i \leq N, 1 \leq j \leq N}$.

Two-party Secure $(A^\tau \cdot B)$ protocol I

1. Alice and Bob jointly generate a $N \times N$ invertible random matrix M;
2. Alice computes $A_1 = A \cdot M_{\text{left}}$, $A_2 = A \cdot M_{\text{right}}$, and sends A_1 to Bob.
3. Bob computes $B_1 = M_{\text{top}}^{-1} \cdot B$, $B_2 = M_{\text{bottom}}^{-1} \cdot B$, and sends B_2 to Alice
4. Alice computes $V_a = A_2 \cdot B_2$ and Bob computes $V_b = A_1 \cdot B_1$.

It can be easily verified that $V_a + V_b = A \cdot B$. Clearly that M plays a central role in the protocol above. A bad choice of M immediately results in a privacy breach. A security analysis of this protocol will be postponed to a later section. First, we look at another protocol for secure matrix product that was proposed by Du et al[21].

Two-party secure $(A^\tau \cdot B)$ protocol II

1. Alice generates a set of $g = \lfloor (N-m)/2 \rfloor$ orthogonal N-dimensional vectors $\{u_1, u_2, \ldots, u_g\}$ such that $u_i^\tau x_j = u_i^\tau z_k = 0$ for all i, j and k. Alice then sends the matrix $U = [u_1, u_2, \ldots, u_g]$ to Bob;
2. Bob computes $W = (I - UU^\tau)B$, where I is an identity matrix, and then sends W to Alice;
3. Alice computes $A^\tau W$, which is equal to $A^\tau B$, and sends it to Bob

Based on any of the protocols above, secure protocols for other more complex matrix operations can be developed. In the following, several of them are briefly described, which are for matrix sum and product, matrix inverse and matrix determinant.

Two-party Secure $((A_1 + B_1) \cdot (A_2 + B_2))$ protocol

In this protocol, A_1 and A_2 are assumed to be held by Alice and B_1 and B_2 by Bob. Observe that

$$(A_1 + B_1) \cdot (A_2 + B_2) = A_1 A_2 + A_1 B_2 + A_2 B_1 + B_1 B_2,$$

To securely compute $(A_1 + B_1) \cdot (A_2 + B_2)$, one only needs to use any of the secure $(A^\tau \cdot B)$ protocols for calculating $A_1 B_2$ and $A_2 B_1$.

Two-party Secure $(A + B)^{-1}$ protocol

In this protocol, Alice owns A while Bob owns B and $A + B$ is invertible. Alice and Bob want to securely compute $(A + B)^{-1}$ such that Alice gets V_a and Bob gets V_b and $V_a + V_b = (A + B)^{-1}$. Again, any of the secure $(A \cdot B)$ protocols can be used to server as the building block. The protocol is stated below.

1. Bob generates two random matrices P and Q;
2. Use one of the secure $(A^t au \cdot B)$ protocols to derive $V_a + V_b = PAQ$. and Alice holds V_a and Bob holds V_b.
3. Bob sends $V_b + PBQ$ to Alice, then Alice compute $V_a + V_b + PBQ = P(A + B)Q$
4. Alice computes $Q^{-1}(A + B)^{-1}P^{-1}$.
5. Using the chosen secure $(A \cdot B)$ protocol further, Alice and Bob drive W_a and W_b respectively such that $W_a + W_b = (A + B)^{-1}$.

Similar techniques can be used to develop secure protocols for calculating matrix determinant $\mid A + B \mid$ and matrix norm $\|A + B\|$. So far, we have developed secure protocols that are effective in computing the regression coefficients while preserving privacy. Recall the formula for $(\hat{\beta}_A, \hat{\beta}_B)$. The procedure is stated in the following protocol.

Secure estimation of regression coefficients protocol

1. Use a secure $(A \cdot B)$ protocol to obtain V_{a1} and V_{b1} such that $V_{a1} + V_{b1} = X^\tau X$;
2. Use the secure $(A + B)^{-1}$ protocol to obtain V_{a2} and V_{b2} such that $V_{a2} + V_{b2} = (V_{a1} + V_{b1})^{-1} = (X^\tau X)^{-1}$.
3. Alice compute $V_{a3} = A^\tau Y$ and $V_{b3} = B^\tau Y$.
4. Use the secure matrix sum-product protocol to compute $\beta = (V_{a2} + V_{b2})(V_{a3} + V_{b3})$.

5.2.2 Diagnostics and Model Determination

As discussed earlier, the estimation of the regression coefficients is the first step in regression. In order to assess whether the model or the fitted model is proper, various diagnostics are required. Most diagnostic procedures rely on residuals, which are the differences between the original responses and the predicted values based on the fitted regression model. Because the fitted regression coefficients are $\hat{\beta}$, so according to Equation 5.4, the predicted or fitted responses are $\hat{Y} = X \cdot \hat{\beta}$ and the residuals are $\hat{\epsilon} = \hat{Y} - Y$. The calculation of $\hat{Y}$ can be carried out securely using the secure sum protocol. Once the residuals are available, some basic diagnostic statistic such as R^2 and the adjusted R^2 can be calculated, which reflect how good the model fits the data. Diagnostic graphics including various residual plots can also be generated, for example, the all-purpose residual versus predicted response plot. If the model is proper and fits the data well, the residual plots are expected to show random clouds of points, or not to show any suspicious patterns. While the residual versus predictor variable plots can tell whether the fitted model is proper, they also can result in violation of privacy. For example, Alice can generate the residual versus x_1 plot, and the coordinates of the points are exactly the values of x_1, if the plot is shown to the other party (Bob). From the plot, Bob may be able to guess accurately the values of x_1 which are held by Alice, thus causing a privacy breach of the attribute values. Some measures can be implemented, so that all the parties can freely share the residual plots to decide if further analyses would be needed, or a acceptable model has already been derived.

The ultimate goal of regression is to obtain a model that best explains the data as well as possesses good accuracy in prediction. Although the model that includes all the attributes has the largest R^2, it is usually a rather poor model because it is highly variable and most likely does not reveal the true relationship between the response and a number of predictor variables. In order to arrive at the best model, model selection is a must in regression. Model selection can be conducted in roughly three modes. It can be an iterative procedure controlled by the analyst based on diagnostic analyses, or an automatic procedure such as stepwise regression, or a exhaustive procedure that runs over all the possible models relying on some model selection criteria such as Mallow's C_p mentioned earlier. At first glance, the secure protocols developed above can be repeatedly used for various models without causing privacy breach. However, after taking diagnostics into consideration, it is not difficult to realize that model selection could present serious challenge to privacy preservation. For example, suppose a model includes only one attribute from each party. Assuming that there are only two parties, the disclosure of the residuals immediately results in the disclosure of the values of the involved attributes to the opposite parties. Unfortunately, this has not been well considered and studied in the literature due to the misunderstanding discussed earlier. Much research is needed to make the secure estimation of regression coefficient practically useful in finding the best regression model while preserving privacy at the same time.

5.2.3 Security Analysis

The two-party secure $(A^\tau \cdot B)$ protocol I and II serve as the fundamental building blocks for the secure estimation of regression coefficients. Both protocols are secure only under the weak secure model. Even so, it is important to understand how secure these two protocols are in the weak sense. Note that both protocols involve the following operation: one party multiplies its data by a $N \times N/2$ matrix known to both parties and sends the result to the other party. Further note that this operation needs to be carried out twice in the two-party secure $(A^\tau \cdot B)$ protocol I, while only once in the two-party secure $(A^\tau \cdot B)$ protocol II. However, following protocol I, in the end, the two parties hold only a portion of $A^\tau B$, while both parties end up knowing $A^\tau B$ following the protocol II. The final result $A^\tau B$ also has implications on privacy disclosure. Hence, deciding which protocol to use depends on the specific application.

This section focuses on the security analysis for the operation described in the previous paragraph. For the ease of presentation, $x = (x_1, x_2, \ldots, x_N)^\tau$ denotes a column of the data matrix and $\tilde{M}$ denote a $N \times N/2$ matrix. Then a party computes $x^\tau \tilde{M}$ and passes the result $b = x^\tau M$, a $N/2 \times 1$ vector, to the other party. If $\tilde{M}$ was not properly chosen, this operation will cause the disclosure of some values of x. For example, if $\tilde{M}$ has a column that is a unit N-dimensional vector with the first coordinate equal to 1 and the other coordinates equal to zero, then the value of x_1 is immediately disclosed, which could lead to serious breach of privacy. In order to prevent this kind of privacy breach from occurring, Du et al. introduced a concept called *k-secure matrix* and carefully studied the properties of these matrices.

Definition 5.1. *Assume* $\tilde{M}$ *is of full rank $N/2$. Let* $\tilde{M}_k$ *be a sub-matrix of* $\tilde{M}$ *by removing k rows from* $\tilde{M}$. $\tilde{M}$ *is k-secure if the rank of* $\tilde{M}_k$ *is* $\frac{N}{2}$ *for any* $\tilde{M}_k$.

The immediate property of k-secure matrices is stated in the following theorem:

Theorem 5.2. *If* $\tilde{M}$ *is k-secure, any nonzero linear combination of the columns of* $\tilde{M}$ *generates a column vector with at least $k + 1$ nonzero entries.*

Let $m_1, m_2, \ldots, m_{\frac{N}{2}}$ be the columns of $\tilde{M}$. Giving $x^\tau \tilde{M}$ is in fact giving the following $\frac{N}{2}$ linear equations: $x^\tau m_1 = b_1, x^\tau m_2 = b_2, \ldots, x^\tau m_{N/2} = b_{N/2}$. Linearly combining these equations will not result in an equation in terms of less than $k + 1$ x_i's. In other words, for any single equation generated as above, there exist at least $k+1$ variables with nonzero coefficients. This further implies that one cannot solve for any of x_i based on the given equations. Any simplified equations involving one particular entry, e.g., x_1, must include another k entries at the same time. In some sense, it can be said that each individual entry is protected by k other entries. For the proof of Theorem 5.2, the reader is referred to Du et al.[21], where the authors also prove other properties of k-secure matrices and argue that $\frac{N}{2}$-secure matrices should be ideal for secure matrix product. Although the larger the value of k, the more secure is the protocol, it is indeed more difficult to construct a k-secure matrix with a large k. Du et al.[21] give the construction of $\frac{N}{2}$-secure matrix using results from coding theory. Similar procedures could be used to generate k-secure matrices for other k. In practice, when the security requirements are not extremely high, lower k-secure matrices can also be used.

By using k-secure matrices, the protocols are secure in a weak sense, but other pitfalls may still exist. For example, although the simplified equations always involve at least $k+1$ variables, some may imply the ranges of some variables if additional information is available. If the same data matrix has been used multiple times and the other party keeps all of the prior results, then these could be combined to infer the individual original values. Overall, the protocols themselves could be regarded to be fairly secure when a large k is enforced. But, as discussed earlier, regression is not just one-step estimation of the linear coefficients, it also involves iterative diagnostics and model selection. Many challenges for preserving privacy exist in these two steps, which have not yet been recognized and studied.

5.2.4 An Alternative: Secure Powell's Algorithm

In the previous sections, all the protocols for secure regression analysis are based on the following formula,

$$\hat{\beta} = (X^\tau X)^{-1} X^\tau Y.$$

As a matter of fact, $\hat{\beta}$ is an explicit minimizer of the following minimization problem

$$\hat{\beta} = \arg\min_{\beta}(Y - X\beta)^\tau (Y - X\beta). \tag{5.9}$$

Recall that the columns of X may be distributed across several parties. For example, when there are two parties, $X = (A, B)$ where A belongs to the first party while B the second party. The protocols discussed in the previous sections tried to secure the direct calculation of $\hat{\beta}$ through Equation 5.3. Recently, Sanil et al. proposed a new method that avoids the direct use of the formula for computing the regression coefficients. Instead, they propose a secure protocol for the minimization of Equation 5.9. As will be discussed later, this leads to higher security for the calculation of $\hat{\beta}$ and may have further implication for secure statistical analysis, and distributed computing, in general. This alternative approach challenges the wisdom of using standard analytical solution in distributed and/or secure computing. The procedure proposed by Sanil et al.[77] is an modification of the well-known Powell's algorithm for quadratic optimization.

Suppose $f(\beta)$ is the target function one wants to minimize, where $\beta \in \mathcal{R}^p$. Powell's algorithm is a derivative-free procedure that finds the minimizer of $f(\beta)$ though a series of line optimizations of $f(\beta)$ in various directions. The following is a pseudo code for the algorithm.

Initialization: Select an arbitrary orthogonal basis for $\mathcal{R}^p$: $d^{(1)}, d^{(2)}, \ldots, d^{(p)}$. Also set an arbitrary starting point $\tilde{\beta}$.

Iteration: Repeat the following block of steps p times

- set $\beta \leftarrow \tilde{\beta}$.
- For $i = 1, 2, \ldots, p$:

  - Find δ that minimizes $f(\beta + \delta d^{(i)})$.
  - Set $\beta \leftarrow \beta + \delta d^{(i)}$.
- For $i = 1, 2, \ldots, (p-1)$: Set $d^{(i)} \leftarrow d^{(i+1)}$.
- Set $d^{(p)} \leftarrow \beta - \tilde{\beta}$.
- - Find δ that minimizes $f(\beta + \delta d^{(p)})$.
  - Set $\tilde{\beta} \leftarrow \beta + \delta d^{(p)})$.

Note that during each iteration in the algorithm above, $(p + 1)$ one-dimensional minimizations need to be carried out. Powell proved that when $f(\beta)$ is a quadratic function, the final output of the algorithm is exactly the minimizer of $f(\beta)$. In regression, the target function $L(\beta) = (Y - X\beta)^\tau (Y - X\beta)$ is clearly a quadratic function, hence the algorithm can be employed directly for computing $\hat{\beta}$. For secure regression, however, additional effort needs to be exerted so that the involved parties do not disclose their attribute values. Sanil et al.[77] assume that all the participants are semi-honest. Utilizing the secure sum protocol, they propose the following procedure for securely computing the regression coefficients.

The key of Powell's algorithm is line optimization. Along any given direction d, the minimizer of $L(\beta + \delta d)$ over δ is

$$\delta = \frac{(Y - X\beta)^\tau X d}{(Xd)^\tau X d} = \frac{z^\tau w}{w^\tau w},$$

where $z = Y - X\beta$ and $w = Xd$. For convenience, we assume that there are three parties denoted by A_1, A_2 and A_3 respectively. The data owned by A_i is denoted by X_{A_i} and the regression coefficients for the attributes held by A_i are denoted by β_{A_i} for $i = 1, 2, 3$. Hence, $X = (X_{A_1}, X_{A_2}, X_{A_3})$ and $\beta^\tau = (\beta_{A_1}^\tau, \beta_{A_2}^\tau, \beta_{A_2}^\tau)$. The given direction d can also be decomposed into three components as $d^\tau = (d_{A_1}^\tau, d_{A_2}^\tau, d_{A_3}^\tau)$. Then,

$$z = Y - [X_{A_1}\beta_{A_1} + X_{A_2}\beta_{A_2} + X_{A_3}\beta_{A_3}]$$

and

$$w = X_{A_1}d_{A_1} + X_{A_2}d_{A_2} + X_{A_3}d_{A_3}.$$

Clearly, z and w can be computed using the secure summation protocol. It is straightforward to extend the discussion above to any number of parties.

The secure Powell's algorithm proposed by Sanil et al.[77] is described as follows. First an initial set of search directions $d^{(1)}, d^{(2)}, \ldots, d^{(p)}$ are chosen such that $d_j^{(r)}$ is zero if r and j are not held by the same party. In the algorithm, each A_j will know and update only $d_{A_j}^{(r)}$, which is the portion related to the attributes owned by A_j. Second, set up the initial values for the regression coefficients $\tilde{\beta}^\tau = (\beta_{A_1}^\tau, \beta_{A_2}^\tau, \ldots, \beta_{A_k}^\tau)$. Third, the following block of steps is iterated p times, which leads to the exact least squares estimate $\hat{\beta}$.

Secure Iteration

1. Each A_j sets $\beta_{A_j} = \tilde{\beta}_{A_j}$.

2. For $r = 1, 2, \ldots, p$:

   (a) Each A_j computes $X_{A_j} \beta_{A_j}$ and $X_{A_j} d_{A_j}^{(r)}$.

   (b) $z = Y - \sum_{j=1}^{k} X_{A_j} \beta_{A_j}$ and $w = \sum_{j=1}^{k} X_{A_j} d_{A_j}^{(r)}$ are computed collectively by $A_1, \ldots, A_k$ using the secure summation protocol.

   (c) Compute $\delta = z^\tau w / w^\tau w$.

   (d) Each A_j updates $\beta_{A_j} \leftarrow \beta_{A_j} + \delta d_{A_j}^{(r)}$.

3. For $r = 1, 2, \ldots, (p-1)$: Each A_j updates $d_{A_j}^{(r)} \leftarrow d_{A_j}^{(r+1)}$.

4. Each A_j updates $d_{A_j}^{(p)} \leftarrow \beta_{A_j} - \tilde{\beta}_{A_j}$.

5. z, w and δ are computed as before, and each A_j updates $\beta_{A_j} \leftarrow \beta_{A_j} + \delta d_{A_j}^{(p)}$.

The secure Powell's algorithm outputs the regression coefficients $\hat{\beta}$ and the residuals which are equal to the final z when the algorithm terminates. Using these results, diagnostics including the all-purpose residual plot and other goodness of fit measures such as R^2 can be generated. Compared to the protocols discussed in Section 5.2.1, the secure Powell's algorithm is at least as competitive in terms of generating the necessary results for fitting a regression model with diagnostics.

Although a rigorous security comparison between the secure Powell's algorithm and the protocols based on Section 5.2.1 would be difficult, it appears that the former ought to be more secure, because, during the iterations of the algorithms, each A_j only updates its portions of the regression coefficients and the search directions. Due to this only some aggregate values of linear combinations of their private data are revealed, while the other parties do not know the coefficients of the linear combinations at all. In other words, the other parties only knows the results at most, do not know the linear equations. In the matrix-product based protocols, the parties do know the linear equations and the values of the these equations, so there exists the risk that they can solve these equations to guess or infer the original values.

A possible downside of the secure Powell's algorithm is that it may not be appropriate for a small number of parties. For example, if there are only two parties, the secure summation protocol will not be able to protect the private data very well. This is true for a small number of parties (> 2) as well, especially when many iterations are needed. Another risk is that, if a party only hold a small number of attributes and it is known to the other parties, the privacy protection it receives is significantly lesser than a party that holds a large number of attributes. For other concerns as well as more details of the technique, readers are referred to Sanil et al.[77].

Most data mining procedures and analyses involve various optimization. This alternative approach to secure regression indicates that non-standard approaches to these optimizations might be more suitable when constraints like privacy protection are present. This idea is worth further investigation in other tasks of statistical analysis and data mining as well.

5.3 Horizontally Partitioned Data

In horizontally partitioned data, each data owner or party holds all the attributes but only a subset of cases. Assume that there are k parties denoted
by P_1, P_2, ..., P_k, who own n_1, n_2, ..., n_k cases, respectively. Let A_i be a
$n_i \times p$ matrix that stores the data of party i, and Y_{A_i} denote the $n_i \times 1$ response vector for the cases held by party i, for $1 \leq i \leq k$. Then the overall
data matrix X and response vector Y have the following representations,

$$
X = \begin{pmatrix} A_1 \\ A_2 \\ \vdots \\ A_k \end{pmatrix} \text{ and } Y = \begin{pmatrix} Y_{A_1} \\ Y_{A_2} \\ \vdots \\ Y_{A_k} \end{pmatrix}.
\tag{5.10}
$$

Each party can use the attribute values (e.g., A_i) and responses (Y_{A_i}) of the
cases it holds to fit a regression model without accessing to other parties'
data. In many applications, however, the parties usually want to fit a regression model collectively due to the following concerns. First, individual parties
may not hold enough data and a model built from individual data may be susceptible to high variability. Second, since the parties own the same attributes
and responses, it would be practically useful to understand how the response
and some of the attributes are related globally assuming a global model exists.
Third, it might also be possible to understand how the relationship between
the response and the attributes changes over the parties. When the parties are
not willing to or not allowed to disclose their data to each other, secure protocols are again needed to facilitate regression for the horizontally distributed
data.

From Equation 5.3 and Equation 5.10, in terms of the distributed data
and responses, the formula for calculating the least squares estimates of β is

$$
\hat{\beta} = (\sum_{r=1}^{k} A_r^\tau A_r)^{-1} (\sum_{r=1}^{k} A_r^\tau Y_{A_r}).
\tag{5.11}
$$

Note that $A_r^\tau A_r$ and $A_r^\tau Y_{A_r}$ can be calculated by P_r locally and confidentially.
Now the secure summation protocol can be employed to calculate the sums in
Equation 5.11, and in the end $\hat{\beta}$ can be obtained. In contrast with vertically
partitioned data, the secure estimation of β for horizontally partitioned data
is relatively much more straightforward.

Once $\hat{\beta}$ is available, the parties can obtain the predicted responses for
their cases, that is, $\hat{Y}_{A_r} = A_r \hat{\beta}$, and further more the corresponding residuals,
$\epsilon_{A_i} = Y_{A_i} - \hat{Y}_{A_i}$. Many diagnostic statistics or measures such as R^2 can be
securely calculated even when the responses are also private, as long as the
statistics are additive across the residuals held by the parties. However, due to
the horizontal distribution of the data, there are privacy concerns regarding
the generation and sharing of various residual plots, which are often fairly

useful diagnostic tools. For example, when the responses are also private, the residual versus predicted responses plot would immediately disclose the response fairly accurately. One possible way to avoid this disclosure is to generate the residual versus predicted response by each party using his/her owned responses and residuals only and then communicate with others about the findings from the plots. However, this naive approach could be easily misleading because it can only reveal local patterns. For the residual versus predictor plots, because the parties hold a subset of the values, an overall plot would again reveal the values of the attributes to all the parties.

In order to facilitate global diagnostics confidentially, Karr et al. [50] proposed to use synthetic predictor values and residuals. The procedure consists of three steps. First, each party simulates values of its predictors. Second, using the estimated $\hat{\beta}$, each party simulates residuals associated with these synthetic predictors in a way that mimics the relationships between the real-data predictors and residuals. Finally, the parties share their synthetic predictors and residuals using secure data integration protocols. The resulting synthetic predictors and residuals then can be used for diagnostics including various residual plots. For detailed description of the procedure, readers are referred to Karr et al. [50]. It is understandable that some information or patterns may be missing in the synthetic data, nonetheless, this is a possible way to conduct diagnostics while protecting original private data.

5.4 Summary and Future Research

Regression is one of the most important and fundamental tasks in data mining and analysis. Unlike other simple tasks such as mining association rules and classification, regression is a complex procedure consisting of at least three crucial components: estimation, diagnostics and model selection. In some sense, regression represents the first real challenge for developing a privacy preserving protocol that can help conduct complex data analyses used in the statistical/actuarial world securely. In the literature, there have been a series of paper attempting to develop protocols and address issues for secure regression, but most of them have only focused on the secure estimation of the regression coefficients of a fixed model. Secure diagnostics have recently received much attention in [50]. Despite these achievements, much effort and research is needed to develop procedures that can accommodate at least the three components effectively as well as securely, so secure regression could be implemented and tested in real applications.

6

Finding Patterns and Rules (Association Rules)

Apart from classification and regression, one of the most important tasks of data mining is to find patterns in data. The best known data analysis technique in this area is association rule mining. In brief, an association rule is an expression of the form $X \Rightarrow Y$, where X and Y are sets of items. The meaning of such rules is quite intuitive – Given a database $\mathcal{D}$ consisting of transactions T (where each transaction is a set of items), a rule of the form $X \Rightarrow Y$ expresses that whenever a transaction T contains X, then it is likely to contain Y. The degree of likeliness is expressed by the confidence of the rule which is defined as the number of transactions containing both X and Y divided by the total number of transactions containing X. That is, the rule confidence is understood as the conditional probability $p(Y \subset T | X \subset T)$. Association rule mining originated from the analysis of market-basket data where rules like "A customer who buys milk and eggs will also buy bread with high probability" are found.

The association rule mining problem can be formally stated as follows[3]: Let $\mathcal{I} = \{i_1, i_2, \cdots, i_m\}$ be a set of literals, called items. Let $\mathcal{D}$ be a set of transactions, where each transaction T is a set of items such that $T \subseteq \mathcal{I}$. Associated with each transaction is a unique identifier, called its *TID*. We say that a transaction T *contains* X, a set of some items in $\mathcal{I}$, if $X \subseteq T$. An *association rule* is an implication of the form $X \Rightarrow Y$, where $X \subset \mathcal{I}$, $Y \subset \mathcal{I}$, and $X \cap Y = \emptyset$. The rule $X \Rightarrow Y$ holds in the transaction set $\mathcal{D}$ with **confidence** c if $c\%$ of transactions in $\mathcal{D}$ that contain X also contain Y. The rule $X \Rightarrow Y$ has **support** s in $\mathcal{D}$ if $s\%$ of the transactions in $\mathcal{D}$ contain $X \cup Y$.

Rules are deemed interesting when they occur in many transactions (high support), and where transactions that contain the left hand side are likely to contain the right hand side (high confidence). The association rule mining problem is to find all such rules, i.e., all A, B, and C that form rules with the desired confidence and support. The key component is to find sets of items that occur frequently, from these the rules can be determined.

The canonical example of association rule mining is that of the original motivators – supermarkets. Any supermarket chain employs market-basket analysis. For example, Wal-Mart analyzes 90 million transactions per week, down to each shopping cart [69]. Indeed, Wal-Mart has been able to find esoteric rules such as: "Wal-Mart customers who purchase Barbie dolls have a 60% likelihood of also purchasing one of three types of candy bars" [69]. In general, rules of this sort can be utilized to reposition items, create new specials, and to generally improve the profitability of a store. When a consumer enrolls in the loyalty/rewards program of the supermarket, the supermarket immediately gets to further mine targeted data for that consumer. This could allow the supermarket to feasibly create specialized coupons for the consumer, and to channel the consumer to useful items. Where does privacy/security play any role in this? The basic model as explained here is certainly very applicable to mining over horizontally partitioned data. Moving data to a central site increases risks of litigation – especially if certain data such as pharmacy data is also moved. Instead of a single Wal-Mart repository, each store can keep ownership over local data, and they can then together mine globally valid rules in a privacy-preserving manner. Also, in order to better compete with these supermarket, local stores which do not have sufficient data on their own can combine data to increase the benefit derived.

6.1 Randomization-based Approaches

Randomization approaches have been proposed for privacy-preserving mining of association rules from horizontally partitioned transaction databases. In real applications, a number of resources can hold a set of transactions. For example, Walmart stores at different locations have their own collections of customer transactions at their stores. These resources want to understand the purchasing behavior of the customers by mining the transaction database collectively, however, they are not willing or allowed to disclose their records to each other or to a third-party analyst. Hence, privacy preserving association rule mining protocols are needed. Because association rules as defined in the introduction are relatively simple patterns and the computation methods for finding association rules such as the popular Apriori algorithm are not complicated, so efficient methods based on secure multi-party computation or cryptography can also be developed, which will be discussed in later sections. In this section, we will focus on the randomization approach. The advantage of randomization approach is that it is usually more efficient than other methods, especially when the databases are of large size. For transaction databases, their sizes are typically in thousands and even millions.

For convenience, we use a simplified model for discussion in the following. Let $\mathcal{I}$ be the collection of items. Suppose there are N clients with transactions $t_1, t_2, \ldots, t_N$, where each transaction t_i is a set of items. The clients need to send their individual transaction to a designated server for mining association rules. Because directly sending the transactions will result in privacy violation, the clients randomize their transactions and send the results to the server instead. The key issues in using the randomization approach are what kind of randomization should be used, how to mine the association rules based on the randomized transactions and how well randomization protects the clients' original transactions. Following [28, 27], we will discuss these issues separately. For simplicity, we assume that all the transactions have the same size, that is, each t_i contains the same number (k) of items. In real applications, the transactions may of different sizes and all the concepts, procedures and results discussed in the following can be modified accordingly.

6.1.1 Randomization Operator

We assume that the transactions are independent and identically distributed as some fixed unknown distribution. A randomization operator, denoted by R, is a random mapping from the realized transactions $(t_i)_{1 \leq i \leq N}$ to all the possible subsets of items from $\mathcal{I}$, i.e.,

$$t_i' = R(t_i), \tag{6.1}$$

where t_i' is a subset of items in $\mathcal{I}$ and $1 \leq i \leq N$.

Note that the definition of randomization operator above is fairly general. Intuitively, the clients do not want to the server to know the items in their original transactions (t_i), so they send the server different transactions (t_i'). If randomization is done systematically, some general association rules and patterns can still be mined from the randomized transactions. This will be discussed in the next subsection. Here, we will introduce several randomization operators. A naïve randomization operator is simply to replace each item in t_i with an item not originally in t_i, with probability p. This operator is called the uniform operator, which can be regarded as a generalization of Warner's "randomized response" method for handling sensitive questions in surveys [87]. Although most items in a transaction can be replaced by using a large p, there is still probability that some items or patterns will be retained, which lead to privacy disclosure. It appears that more sophisticated randomization operators are needed. [28] proposed an operator called "select-a-size" randomization defined as follows.

Definition 6.1. *Let ρ be a fixed number between 0 and 1, and let $\{p[j]\}_{0 \leq j \leq m}$ be a probability distribution over $\{0, 1, 2, \ldots, m\}$. Given a transaction, the select-a-size randomization operator generates another transaction $t' = R(t)$ in three steps:*

1. *The operator randomly draw an integer j from $\{0, 1, 2, \ldots, m\}$ using the distribution $\{p[j]\}_{0 \leq j \leq m}$.*
2. *It selects j items from t uniformly at random without replacement. These items, and no other items of t, are placed into t'.*
3. *It considers each item $a \notin t$ in turn and tosses a coin with probability ρ of "heads" and $1 - \rho$ of "tails". At those items for which the coin faces "heads" are added to t'.*

Both the uniform and the select-a-size operators are per-transaction and item-invariant, because they apply the same randomization procedure to the transactions independently and do not use any item specific information. Compared to the uniform operator, the select-a-size operator tries to include more false items into t', leading to randomized transactions that better hide the original ones. [28] also described a special case of the select-a-size operator called the cut-and-paste randomization operator, which has fewer parameters and is relatively easier to implement.

6.1.2 Support Estimation and Algorithm

Once the server has received the randomized transactions, it need to mine the original association rules from these data. The key in ordinary association rule mining is to find subsets of items that have support above a prespecified threshold. In this section, we introduce a method to estimate the original support of a given subset of items based on the randomized data and the associated mining algorithm.

Let $T = \{t_i\}$ be the original transactions and $T' = \{t'_i\}$ the randomized transactions. Let A be a fixed subset of items of size k. The support of A in the original transactions T is denoted by s and in the randomized transaction by s'. To derive the estimate of s, a concept called partial support of A is defined as follows. For $0 \leq l \leq k$,

$$s_l = \frac{\#\{t \in T \mid \#(A \cap t) = l\}}{N} \tag{6.2}$$

where $\#()$ is the cardinality or size of a set. s_l is called the partial support of A for intersection size l in T. Clearly $s_k = s$ because the size of A is k. Similarly, partial support of A for intersection size l in T' can be defined and is denoted by s'_l for $0 \leq l \leq k$. Furthermore, define the *transitional probability from intersection size l to intersection size l'* as

$$p[l \rightarrow l'] = P[\#(t' \cap A) = l' \mid \#(t \cap A) = l] \tag{6.3}$$

for $0 \leq l, l' \leq k$. Because the randomization operator is pre-transaction and item-invariant, $p[l \rightarrow l']$ does not depend on the transaction t, in fact, only depends on m, k, l l', ρ and the distribution $\{p[j]\}_{0 \leq j \leq m}$. Hence when the parameters are given, $p[l \rightarrow l']$ can be calculated; the explicit formula for $p[l \rightarrow l']$ can be found in [28]. Let $P = (P_{l',l})$ be the $(k+1) \times (k+1)$ matrix with entries $P_{l',l} = p[l \rightarrow l']$.

Now we are ready to state an important result derived by [28], which relates the partial supports of A in the randomized transactions T' and the partial supports of A in the original transactions T. Note that, due to the select-a-size randomization operator, the partial supports $s'_0, s'_1, \ldots, s'_k$ are random variables. [28] showed that the expected values of $s'_0, s'_1, \ldots, s'_k$ are given by

$$E(s'_0, s'_1, \ldots, s'_k)^\tau = P(s_0, s_1, \ldots, s_k)^\tau, \tag{6.4}$$

where P is the matrix of transitional probabilities defined above. Now let $Q = (Q_{l,l'})$ be the inverse matrix of P when it exists. Then

$$(s_0, s_1, \ldots, s_k)^\tau = QE(s_0, s'_1, \ldots, s'_k). \tag{6.5}$$

Therefore,

$$s_k = E(Q_{k,0}s'_0 + Q_{k,1}s'_1 + \cdots + Q_{k,k}s'_k). \tag{6.6}$$

Define

$$\hat{s} = Q_{k,0}s'_0 + Q_{k,1}s'_1 + \cdots + Q_{k,k}s'_k. \tag{6.7}$$

Equation 6.5 indicates that $\hat{s}$ is an unbiased estimate of s_k, which is exactly the support (s) of A in T. In other words, from the randomized transactions, we can calculate all the partial supports of A, which can be used to derive an unbiased estimate of s. The variance of $\hat{s}$ and its estimate were also derived explicitly in [28], which can be used to assess the accuracy of $\hat{s}$. Readers are referred to [28] for the details.

Most algorithms used for mining association rules in T such as the Apriori algorithm can be modified to mine association rules from T'. First, all the partial supports of a candidate subset need to be calculated in T' in order to estimate s. Second, the criteria to retain or discard candidate matrices are still in terms of s instead of $s'_0, s'_1, \ldots, s'_k$, so some special treatments are needed. Readers can find a modified Apriori algorithm for mining association rules from T' in [28].

6.1.3 Limiting Privacy Breach

Intuitively, the select-a-size randomization operator is expected to protect the items in original transactions and preserve the privacy of the clients on average. However, for each individual client, a type of privacy violation called privacy breach can still occur. [27] carefully defined privacy breach and proposed a method called amplification that can be used to limit privacy breaches. The concept and method are generally applicable to privacy preserving data mining via randomization. Hence, following [27], we introduce them in general, then apply them to mining association rule using the select-a-size randomization operator.

In general, we assume N clients hold private data $x_1, x_2, \ldots, x_N$, respectively, which can be regarded as realizations of a random variable X with distribution $p_X(x) = P[X = x]$ and sample space V_x. Using a randomization operator R, each client sends the randomized value $y_i = R(x_i)$ to a server for analysis. The randomized data $y_1, y_2, \ldots, y_N$ are regarded as realizations of $Y = R(X)$ with distribution $p_Y(y) = P[Y = y]$ and sample space V_Y. When the randomized data y_i is disclosed, the posterior probability for the corresponding client's original private data to be $x \in V_X$ is

$$P[X = x \mid Y = y_i] = \frac{P[X = x] \cdot p[x \to y_i]}{P[Y = y_i]}, \qquad (6.8)$$

where $p[x \to y_i]$ is the transitional probability for R to map x to y_i; and the posterior probability that the client's data possesses a property $C(X)$ is

$$P[C(X) \mid Y = y_i] = \sum_{Q(x), x \in V_X} P[X = x \mid Y = y_i] \qquad (6.9)$$

where $C : V_X \to \{\text{true}, \text{false}\}$. For example, $C(X)$ is true when X belongs to a particular subset. Informally, the disclosure of y_i significantly increases the probability that the server knows that the client possesses a property $C(X)$. Sometimes, the lack of a certain property is considered to be private. so if the disclosure of y_i significantly decreases the probability that a client possesses a property, it also results in privacy breach. Given below are the formal definitions of these two types of privacy breach.

Definition 6.2. *Let C_1 and C_2 be two arbitrary properties, and let $0 < \rho_1 \leq \rho_2 < 1$. It is said that there is a upward $\rho_1-to-rho_2$ privacy breach with respect to C_1 if for some $y \in V_Y$*

$$P[C_1(X)] \leq \rho_1, \ P[C_1(X)|Y = y] \geq \rho_2; \qquad (6.10)$$

it is said that there is a downward $\rho_2-to-rho_1$ privacy breach with respect to C_2 if for some $y \in V_y$

$$P[C_2(X)] \geq \rho_2, \ P[C_2(X)|Y = y] \leq \rho_1. \qquad (6.11)$$

Randomization operators need to be designed so that they will not lead to either type of privacy breach defined above. Directly verifying whether a given operator causes privacy breach or not is generally tedious or infeasible. Fortunately sufficient conditions exist, which guarantee neither of the two types of breach will happen for given ρ_1 and ρ_2. One such condition called amplification was given by [27] in terms of the transitional probabilities $p[x \to y]$. The intuitive idea behind the condition is that, if the probabilities for the values of X to be randomized to y are close to each other, the disclosure of $y = R(x)$ does not tell much about the original value x. In other words, the transitional probability $p[x \to y]$ should not be too large compared to the other values of X for any fixed x and y.

Definition 6.3. *A randomization operator $R(x)$ is at most $\gamma-$amplifying for $y \in V_y$ if*

$$\frac{p[x_1 \to y]}{p[x_2 \to y]} \leq \gamma \qquad (6.12)$$

for any x_1 and x_2 in V_X such that $p[x_2 \to y] > 0$ where $\gamma \geq 1$. Operator $R(x)$ is at most $\gamma-$amplifying if it is at most $\gamma-$amplifying for all suitable $y \in V_Y$.

Let $y \in V_y$ and assume there exists $x \in V_X$ such that $p[x \to y] > 0$. [27] showed a sufficient condition for $R(x)$ to cause neither the upward $\rho_1 - \text{to} - \rho_2$ privacy breach nor the downward $\rho_2 - \text{to} - \rho_1$ privacy breach is that R is at most γ−amplifying where γ satisfies

$$\frac{\rho_2}{\rho_1} \frac{1 - \rho_1}{1 - \rho_2} > \gamma. \tag{6.13}$$

Next we apply the general results to the select-a-size randomization operator for mining associations rules. Under the assumptions in Sections 6.1.1–6.1.2, V_X consists of transactions of size m and V_Y of all the possible transactions. Let n be the number of items in $\mathcal{I}$. Let t be an arbitrary transaction in V_X and $t' = R(t)$. Let $m' = \#(t')$, i.e., the number of items in t' and $j = \#(t \cap t')$. The transitional probabilities from t to t' is

$$p[t \to t'] = \frac{p[j]}{\binom{m}{j}} \rho^{m'-j} (1 - \rho)^{n-m-m'+j}, \tag{6.14}$$

where $\{p[j]\}$ is the distribution used to select a size for randomization. For any two transactions, say t_1 and t_2, in V_X such that $\#(t_1 \cap t') = j_1$ and $\#(t_2 \cap t') = j_2$,

$$\frac{p[t_1 \to t']}{p[t_2 \to t']} = \frac{p[j_1]}{p[j_2]} \frac{\binom{m}{j_2} \rho^{j_2} (1 - \rho)^{m-j_2}}{\binom{m}{j_1} \rho^{j_1} (1 - \rho)^{m-j_1}} \tag{6.15}$$

Note that $\frac{p[t_1 \to t']}{p[t_2 \to t']}$ depends on j_1, j_2, m and the tuning parameters ρ and $\{p[j]\}$ of the select-a-size randomization operator R. Applying Equation 6.13 the sufficient condition for R to cause neither the upward ρ_1-to-ρ_2 privacy breach nor the downward ρ_2-to-ρ_1 privacy breach is

$$\frac{p[j_1]}{p[j_2]} \frac{\binom{m}{j_2} \rho^{j_2} (1 - \rho)^{m-j_2}}{\binom{m}{j_1} \rho^{j_1} (1 - \rho)^{m-j_1}} \leq \gamma \tag{6.16}$$

for $0 \leq j_1, j_2 \leq m$ where γ satisfies Equation 6.13. Hence, any select-a-size operator satisfying the sufficient condition can prevent the two types of privacy breaches from occurring. In order to choose the best possible operator, [27] further considered the amount of aggregate information that can be transmitted to the randomized transactions and proposed a simple but optimal way to determine the tuning parameters in R. Other properties of R have also been studied in [27]. Readers are referred to the original works for more details and applications in real databases.

6.1.4 Other work

One limitation of the current randomization based approaches for association rule mining is that the randomization is both *transaction invariant* as well as *item invariant*. Under the assumption that each data provider holds only a single transaction, while the data mining is done by a single central site, Zhang, Wang and Zhao [91] propose a scheme based on algebraic techniques. However, their technique is limited to semi-honest adversaries.

The key difference is that perturbation guidance is provided to the data provider. This is done by dynamically estimating the eigenvectors of the transaction matrix. Since transactions are provided to the central site in a stream fashion, the transaction matrix is updated by appending each new transaction at the end. After update, singular value decomposition gives the eigenvectors and eigenvalues of the transaction matrix. The k eigenvectors corresponding to the top k eigenvalues form the perturbation guidance V_k given to the next data provider. Thus, V_k itself is a $n \times k$ matrix. Given V_k, each data provider applies the perturbation function $R(\cdot)$ to its data transaction t. $R(t)$ is defined as follows: first the data transaction t in vector form (boolean) is transformed to the integer vector $\tilde{t} = tV_kV_k'$. This integer vector is then mapped back to a boolean vector using a very simple transformation utilizing a pre-defined parameter ρ_t. If $\tilde{t}_i \geq 1 - \rho_t$, the corresponding element $R(t)_i \leftarrow 1$ otherwise $R(t)_i \leftarrow 0$. Finally, the transformed transaction $R(t)$ is additionally perturbed as follows: for every item not in the original transaction, a random number r is chosen from a uniform distribution over $[0, 1]$. If $r \geq 1 - \rho_m$ where rho_m is another pre-defined parameter, then the item is added to $R(t)$. This completes the transformation of a transaction.

[91] further analyze the accuracy and privacy properties of this algorithm and compare it to the cut and paste operator proposed for randomization by Evfimievski et al[28].

Rizvi and Haritsa [75] also propose a data distortion based approach to mine association rules from boolean data. Again, the idea is to modify data values such that reconstruction of the values for any individual transaction is difficult, but the rules learned on the distorted data are still valid. One interesting feature of this work is a flexible definition of privacy; e.g., the ability to correctly guess a value of '1' from the distorted data can be considered a greater threat to privacy than correctly learning a '0'. The basic distortion procedure is as follows: The distorted vector Y is generated from the boolean source vector X as follows: $Y_i = X_i \oplus \bar{r}_i$, where $\bar{r}_i$ is the complement of r_i, a random variable with density function $f(r) = bernoulli(p)(0 \leq p \leq 1)$. Thus, r_i takes a value 1 with probability p and 0 with probability $1 - p$. The net effect of the above computation is that the identity of the ith element in X is kept the same with probability p and is flipped with probability $(1 - p)$. This distorted vector is provided to the database miner. The distorted database can be mined provided that the distortion procedure as well as the value of p

is known to the miner. For further details, readers are referred to the paper by Rizvi and Haritsa [75].

6.2 Cryptography-based Approaches

Several cryptography based approaches have also been proposed to solve the association rule mining problem. We introduce two techniques here, which show the different challenges (and solutions) resulting from horizontally and vertically partitioned data.

6.2.1 Horizontally Partitioned Data

In a horizontally partitioned database, since the transactions are distributed among the k sites, the global support count of an item set is the *sum* of all the local support counts. Thus, an itemset X is *globally supported* if the global support count of X is bigger than $s\%$ of the total transaction database size. The global confidence of a rule $X \Rightarrow Y$ can be given as $\{X \cup Y\}.sup/X.sup$. An itemset is called a globally large itemset if it is globally supported.

The aim of distributed association rule mining is to find all rules whose global support and global confidence are higher than the user specified minimum support and confidence.

Kantarcioglu and Clifton[47] propose a secure method based on the FDM algorithm[13]. The FDM algorithm is a fast method for distributed mining of association rules as summarized below:

1. **Candidate Set Generation**: Intersect the globally large itemsets of size $p-1$ with locally large $p-1$ itemsets to get candidates. From these, use the classic apriori candidate generation algorithm to get the candidate p itemsets.
2. **Local Pruning**: For each X in the local candidate set, scan the local database to compute the local support of X. If X is locally large, it is included in the locally large itemset list.
3. **Itemset Exchange**: Broadcast locally large itemsets to all sites – the *union* of locally large itemsets, a superset of the possible global frequent itemsets. (It is clear that if X is supported globally, it will be supported at least at one site.) Each site computes (using apriori) the support of items in union of the locally large itemsets.
4. **Support Count Exchange**: Broadcast the computed supports. From these, each site computes globally large p-itemsets.

The FDM algorithm as described above avoids disclosing individual transactions, but does expose significant information about the rules supported at each site. The goal is to approximate the efficiency of the above algorithm, without requiring that any site disclose its locally large itemsets, support counts or transaction sizes.

The Kantarcioglu and Clifton algorithm[47] basically modifies the above outlined method. In the **Itemset Exchange** step, a secure union algorithm is used to get the global candidate set. After this step, the globally supported itemsets can be easily found using a protocol for secure sum. The output of the secure sum protocol is the actual sum. However, rather than learning the exact support, what is really required is to simply determine whether the support exceeds a threshold. To do this without revealing the actual sum, the secure sum algorithm is modified slightly. Instead of sending $R + \sum v_i$ to site 1, site k performs a secure comparison with site 1 to see if $R + \sum v_i \geq R$ (using a circuit, as described in Chapter 3.3.1). If so, the support threshold is met. The confidence of large itemsets can also be found using this method. One thing that needs to be emphasized is that if the goal is to have a totally secure method, the union step itself has to be eliminated. However, using the secure union method gives high efficiency with provably controlled disclosure of some minor information (i.e., the number of duplicate items and the candidate sets.) The validity of even this disclosed information can be reduced by noise addition. Basically, each site can add some fake large itemsets to its actual locally large itemsets. In the pruning phase, the fake items will be eliminated.

This gives a brief, oversimplified idea of how the method works. Full discussion can be found in [46].

6.2.2 Vertically Partitioned Data

For vertically partitioned data, a secure algorithm can be created simply by extending the existing apriori algorithm. Remember that vertical partitioning implies that an itemset could be split between multiple sites. Most steps of the apriori algorithm can be done locally at each of the sites. The crucial step involves finding the support count of an itemset. If the support count of an itemset can be securely computed, one can check if the support is greater than threshold, and decide whether the itemset is frequent. Using this, association rules can be easily mined securely.

Now, we look at how this can be done. Consider the entire transaction database to be a boolean matrix where 1 represents the presence of that item (column) in that transaction (row), while 0 correspondingly represents an absence. The key insight is as follows: The support count of an itemset is *exactly* the scalar product of the vectors representing the sub-itemsets with both parties. Thus, if we can compute the scalar product securely, we can compute the support count. Full details are given in [80] which proposes an algebraic method to compute the scalar product. While this algebraic method is not provably secure, other methods since then have been proposed for computing the scalar product[20, 43, 35], out of which at least one [35] is provably secure.

These protocols typically assume a semi-honest model, where the parties involved will honestly follow the protocol but can later try to infer additional information from whatever data they receive through the protocol. One result of this is that parties are not allowed to give spurious input to the protocol.

If a party is allowed to give spurious input, they can probe to determine the value of a specific item at other parties. For example, if a party gives the input $(0, \ldots, 0, 1, 0, \ldots, 0)$, the result of the scalar product (1 or 0) tells the malicious party if the other party the transaction corresponding to the 1. Attacks of this type can be termed probing attacks and need to be protected against.

Another way of finding the support count is as follows: Let party i represent its sub-itemset as a set S_i which contains only those transactions which contain the sub-itemset. Then the size of the intersection set of all these local sets ($|S|, S = \cap_{i=1}^{k} S_i$), gives the support count of the itemset. The key protocol required here is a secure protocol for computing the size of the intersection set of local sets.

Evfimievski et al[2] and Vaidya and Clifton[84] describe two similar protocols for doing this based on commutative encryption. Freedman et al. [33] also propose techniques using homomorphic encryption to do private matching and set intersection for two parties which can guard against malicious adversaries in the random oracle model as well. We now briefly describe the technique in [84]. Commutative encryption means that the order of encryption does not matter. Thus if an item X is encrypted first with a key K_1 and then with a key K_2, the resulting ciphertext is the same if X was encrypted first with K_2 and then with K_1. Thus, $Y = E_{K_2}(E_{K_1}(X)) = E_{K_1}(E_{K_2}(X))$. What makes this useful is that K_1, K_2 can be generated by separate parties and kept secret. As long as the keys are not known to other parties, the ciphertext cannot be decrypted, and it is impossible to know what has been encrypted. Thus, every party i generates its own key K_i, and every party encrypts the local set of each party with its key. At the end, all the sets have been encrypted by all of the keys, and the common items in the sets have the same ciphertext – thus, one can easily count the size of the intersection set without knowing what the common items are. This is one of the protocols proposed in [2] by Evfimievski et al. However, the protocol is restricted to only two parties, assumes semi-honest adversaries and is susceptible to the probing attacks described earlier. The protocol proposed by Vaidya and Clifton [84] utilizes the same basic idea but has some enhancements so that it can applied to any number of parties and provides resistance to probing attacks.

For finding the support of a single itemset, either of the two ways presented in the earlier couple of paragraphs is equally good. However, for association rule mining, both are not quite the same. The second method (of set intersection) is much more efficient than the first (of scalar product). The reason is rather subtle. Basically, in set intersection using commutative encryption, the protocol actually leaks more information than purely necessary. Instead of simply finding out the size of the complete intersection set, each party can also compute the sizes of the intermediate intersection sets (of a subset of the attributes). This fact can be turned to our advantage. Now, all parties simply encrypt all of the attributes with their keys. Once all attributes have been encrypted with all of the keys, they are sent to all of the parties. At this point, every party can simply run Apriori locally and the do the necessary intersections to find out the support of an itemset. Thus, the overall computation/communication complexity is independent of the number of itemsets (which can run in the thousands).

6.3 Inference from Results

The techniques described above all provide a measure of protection for the data from which the rules are derived. The rules, by design, give knowledge. The problem is that this knowledge can inherently pose a risk to privacy. For example, assume we are performing a medical study to determine risk factors for a terminal disease. The methods described earlier in this chapter ensure that individuals having the disease are not disclosed. However, a high confidence rule, combined with external knowledge about individuals, could enable inference about an individual's disease status. Specifically, if the antecedents of the rule are not private information, a high confidence rule inherently discloses that individuals who meet the rule criteria are likely to have the disease.

Addressing this problem is challenging, as the very outcome we desire (association rules) is exactly the source of the privacy violation. Quantifying the probability of disclosure of private information is straightforward; confidence gives a probability distribution on the private value for individuals meeting the antecedent, support says what fraction of individuals will meet that antecedent. Combining this with an estimate of the likelihood that an adversary will know the attributes in the antecedent of the rule for a given individual allows an estimate of the privacy risk to that individual.

More difficult is deciding if the risk posed is acceptable, particularly in light of the value of the knowledge in the rule. While Chapter 2 gives some insight into this issue, there is still research needed to build a comprehensive framework for evaluating the value of privacy.

If the risk to privacy outweighs the reward, does this eliminate our ability to mine data? In [5], Atallah et al. pose an interesting alternative: altering data to lower the support and/or confidence of a specific rule. In many ways, this is similar to the randomization of Section 6.1 – transactions are altered by inserting "fake" items or removing real items. The goal, however, is to remove specific rules from the output rather than protect the input data.

The problem of minimizing the changes to the data while still ensuring that the confidence and support of the rules in question are reduced below a threshold was shown to be NP-hard in [5]. The paper proceeded to give a heuristic algorithm, and demonstrate bounds on the amount of data modification needed. This was later extended to adding unknown values in [78] (although this only works if the data has a significant number of unknown values to begin with, otherwise the unknown values can be used to identify what rules might have been hidden.)

We do not go into detail on these approaches, because they do not really address privacy in any common sense of the term. A complete treatment of this topic can be found in [85]. In conjunction with randomization based approaches, these methods may be effective at addressing both privacy and the inference problem posed at the beginning of this section, providing a more complete solution for privacy-preserving association rule mining. This is a challenging and open problem for future work, and probably requires significant advances in privacy definitions and metrics to come to full fruition.

A first cut at a full evaluation of the potential inference problems from association rules was given by Mielikäinen [24]. He formulated an *inverse frequent set mining* problem: Given a set of association rules, what data sets are compatible with that set of rules? While perhaps not immediately apparent, this is actually an effective metric for privacy. By measuring the fraction of such compatible data sets that disclose an individual's private data, the risk of disclosure posed by the rules can be estimated. If the adversary is presumed to know certain external information, the data sets can be further limited to those consistent with the external information, and again the disclosure risk can be measured by the fraction that correctly identify an individual's private data.

Unfortunately, the value of this approach to estimating privacy risk is tempered by the fact that inverse frequent set mining is NP-complete; this was shown through equivalence to 3-coloring in [24]. While this may also seem fortunate, as it would seem to make the adversary's job difficult, relying on NP-hardness to ensure security has often failed in areas such as cryptography. In particular, [24] also shows that for some sets of frequent itemsets, determining the compatible datasets is polynomially computable.

While this section has been a slight foray from the task of computing data mining models without revealing the data used, it has pointed out the challenge of ensuring that private *knowledge* is not revealed. Even when individually identifiable data is protected, the knowledge gained through data mining can pose a privacy risk. Evaluating the risk, and the tradeoff between the risk and reward from obtaining that knowledge, is still an area of active research. This will be discussed further when we look at open privacy-preserving data mining challenges in Chapter 8.

7

Descriptive Modeling (Clustering, Outlier Detection)

The data mining tasks of the previous chapters – classification, regression, associations – have a clearly defined "right answer". While it may not be possible to learn that "right answer" (e.g., we may not develop an optimal Bayesian classifier), the algorithms follow clearly defined paths. Descriptive modeling is less clearly defined. With clustering, for example, not only do we not know in advance what the clusters mean, we may not even know the proper number of clusters. Descriptive modeling is a much more *exploratory* process.

This poses new challenges for privacy-preserving data mining. Algorithms for descriptive modeling tend to be iterative. The simple act of tracking data access across iterations can be revealing. For example, if an outlier detection algorithm frequently access a particular individual, it may imply that the individual is actually quite central. This is because algorithms that detect outliers by noting a lack of near neighbors would frequently access a central individual when attempting to show that other "near-center" entities are *not* outliers. Contrast with something like association rules, where the data items accessed at each iteration can be determined knowing the rules produced by that iteration.

Perhaps as a result of this distinction, there has been little work on perturbation-based approaches to privacy-preserving clustering. If the goal of clustering were to model the clusters, running algorithms directly on perturbed data may well give reasonable results. However, if the goal is to determine the cluster that an individual belongs to (or if the individual is an outlier), perturbation-based techniques will give completely distorted results – even though the general clusters may be okay, *which* individual is in which cluster would be completely altered.

The challenges are high for other approaches, as well. Iteration poses a challenge for secure multiparty computation, in that the number of iterations or intermediate steps may reveal information that compromises privacy (e.g., an item moving between clusters as cluster centers move gives more information about the values of that item than simply which cluster it is in.) In this chapter we look at three approaches to clustering. The first introduces a new approach, *data transformation*. As with perturbation, the goal is to protect privacy by altering the data. The difference is that the transformation approach preserves the information needed for clustering, while eliminating the original (private) data values. The other two approaches are based on the secure multiparty computation model, and show how iteration challenges can be addressed. We finish with a privacy-preserving approach to the closely related problem of outlier detection.

7.1 Clustering

Several solutions have been proposed for privacy-preserving clustering. One question that arises is why the parties need to collaborate in the first place? Would not some form of local clustering with model merging suffice? However, it can be easily seen that global data does give significantly better descriptive models. Figure 7.1 shows a simple two-dimensional plot of data points clustered into 3 clusters. If we consider the data to be vertically partitioned, one party would know the projection on the X-axis, while the other would know the projection on the Y-axis. Running a clustering algorithm such as k-means locally would give poor results.

Figure 7.2 shows the clusters that would probably be found by the first party. Figure 7.3 shows the clusters detected by the other party. It is easy to see that these are incorrect, and furthermore, difficult to combine into a valid model. Indeed from x's point of view (looking solely at the horizontal axis), there really are only two clusters, "left" and "right", with both having a mean in the y dimension of about 3. Thus, a method such as X-means [70] would probably stop after finding these two clusters. The problem is exacerbated by higher dimensionality and clearly shows the need for global clustering.

7.1.1 Data Perturbation for Clustering

Disclosing distorted data instead of the original data is a natural way to protect privacy. As discussed in the previous chapters, randomization / perturbation randomly modifies individual data and the resulting data can be mined for association rules, regression analysis etc. Though randomization should still work for privacy-preserving clustering, more effort will be needed. Furthermore, when the clusters in the original data are not well separated from each other, it may be impossible to identify them in the randomized data. Furthermore, even if appropriate clusters are identified, the proper cluster for each individual is likely to be lost. It is easy to imagine scenarios where private data would be used to group individuals, with the goal of identifying which individuals belong in which group rather than statistics on the clusters (computer dating is one thought.)

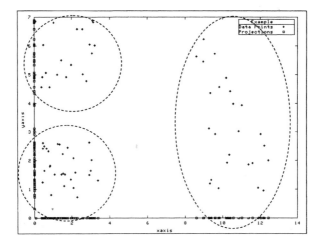

Fig. 7.1. A two-dimensional scatter plot grouped into 3 clusters

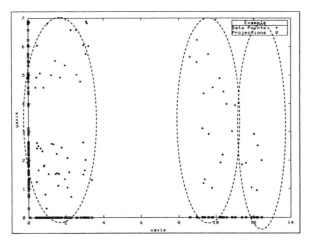

Fig. 7.2. Clusters found by the first party on local data

Transformation-based Clustering

To address this problem, while maintaining the characteristics of the perturbation approach, Oliveira and Zaïane proposed to use transformation, instead of randomization, to distort the original data [66, 67].

Clustering is to divide data points into groups, so that the points in the same group are similar to each other but the points from different groups are dissimilar with each other. Hence, the measure of similarity or dissimilarity between two data points play a crucial rule in clustering. Different measures lead to different clustering methods. [66] regarded data points as vectors in multiple dimensional Euclidean space and used the Euclidean distance as the similarity measure. Let $D = (d_{ij})_{m \times n}$ be an $m \times n$ data matrix in which each

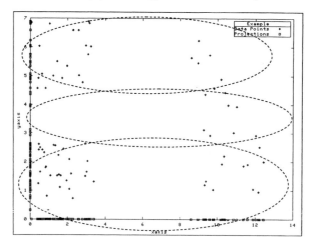

Fig. 7.3. Clusters found by the second party on local data

row corresponds to a data point. Assume that D is held by an owner that requests another party to conduct a clustering analysis of D but do not want to disclose D. The owner decides to transform the data points and then send the transformed data denoted by D' to the other party for analysis. Because the clustering result based on D' should be exactly the same as that based on D, the similarity between any two rows of D should be invariant under the transformation. This is guaranteed by using orthogonal transformations. In [66], in fact, only a subfamily of orthogonal transformation was proposed.

Let D_i denote the ith column of D for $1 \leq i \leq n$. Let

$$R = \begin{bmatrix} \cos \theta & \sin \theta \\ -\sin \theta & \cos \theta \end{bmatrix} \qquad (7.1)$$

where θ is an angle. R is a two-dimensional rotation operator. Any two given columns of D, D_i and D_j, can be rotated as

$$(D_i', D_j') = (D_i, D_j)R. \qquad (7.2)$$

Replacing D_i and D_j with D_i' and D_j' leads to a new data matrix where the ith and jth columns have been modified. This operation can continue until all the columns of D have been modified, and the resulting matrix D' has the property that the distances between the rows of D' are equal to the distances between the rows of D. Note that if θ is 0, then R is an identity matrix and $D_i' = D_i$ and $D_j' = D_j$. In other words, the columns have not been really modified. Therefore, θ needs to be carefully chosen so that the modified columns are different from the original ones. [66] proposed the pairwise-security thresholds ρ_1 and ρ_2 for selecting θ, which are

$$\text{Var}(D_i - D_i') \geq \rho_1 \text{ and } \text{Var}(D_j - D_j') \geq \rho_2, \qquad (7.3)$$

where Var() is the variance of the vector entries. Hence, θ should be chosen so that the thresholds above are satisfied. The above procedure is repeated for a number of disjoint or joint pairs of the columns until all the columns of D are distorted. This method was referred to as Rotation-Based Transformation (RBT) in [66]. Following is the pseudocode for RBT.

RBT.Algorithm

Input: $D_{m \times n}$ **Output:** $D'_{m \times n}$

1. Choose k pairs of security thresholds ($k = \lceil \frac{n}{2} \rceil$). $\{(\rho_{l1}, \rho_{l2})\}_{1 \leq l \leq k}$.
2. Choose k pairs of columns of D, e.g., (D_i, D_j), such that each column of D is selected at least once.
3. **For each** selected pair (D_i, D_j) **do**
   a) Compute $(D_i, D_j)R(\theta)$ as a function of θ.
   b) Select θ_l such that $\mathrm{Var}(D_i - D'_i) \geq \rho_{l1}$ an $\mathrm{Var}(D_j - D'_j) \geq \rho_{l2}$
   c) $(D'_i, D'_j) = (D_i, D_j)R(\theta_l)$.

End.for

End.Algorithm

   Although the RBT procedure was proposed for centralized data, it can be extended to distributed or partitioned data without much difficulty. To generalize it to similarity measures other than the Euclidean distance would be difficult. In addition, in practice, one may know ahead of time which clustering methods and which similarity measure will be used. This drawback could limit the usefulness of this procedure. Another concern is the security of the procedure. [66] argued that a brute force attack would require a great deal of computational power to get the original data, because an adversary has to figure out the correspondence between the pairs and the angles θ_l used during the transformation. The procedure outlined above only require every column of D to be rotated at least once, which may not be enough to hide the data well. We believe that for data of moderate size, the computation to recover the original data or some columns of D may not as intensive as [66] thought. Furthermore, once a privacy breach occurs, it is not just limited to one individual, in fact, it will cause the disclosure of all the values of at least one attribute. Likewise, prior knowledge of the attributes of a few individuals (as many as the number of attributes) enables triangulation using the distances alone to determine the entire data; inverting the rotation is not necessary. Hence, the security of this kind of transformation is also questionable in practice. Transformation is a valuable idea, but more sophisticated procedures need to be developed.

Generative Models for Clustering

As discussed earlier, disclosing summary statistics instead of original data is an effective way to share information and preserve privacy. The procedure for privacy preserving clustering using generative models [66] belongs to this category. Suppose data are horizontally distributed among k sources. [66]

assumes that there exists an underlying distribution that generates the data and it can learned from high-level information from the k partitioned data instead of integrating the data together. Under this assumption, a source does not need to send actual data to a third party for clustering, instead, a model is fitted to the local data, then the model is sent to a server to learn the global distribution.

Let the k resources be labeled by S_1, S_2, ..., S_k, respectively. At S_i, model-based clustering methods are employed to fit a model denoted by p_i. Because the disclosure of a model may also cause a privacy breach, certain privacy cost constraints need to be satisfied by p_i. In [66], the authors advocated the use of likelihood of the data instead of entropy to quantify privacy loss. Then the sources send the models $\{p_i\}_{1 \leq i \leq k}$ to a centralized server that learns a global model for clustering. [66] uses KL-divergence to measure the discrepancy between two densities f and g, which is

$$K(f;g) = \int f(x) \log \frac{f(x)}{g(x)} dx. \tag{7.4}$$

The global model p_0 can be approximated by the solution of the following optimization problem:

$$p^* = \mathrm{argmin}_{p \in \mathcal{F}} \sum_{i=1}^{k} \nu_i K(p_i; p), \tag{7.5}$$

where $\mathcal{F}$ is a family of distributions, and $\{\nu_i\}_{1 \leq i \leq k}$ are the proportions of the data held by $\{S_i\}_{1 \leq i \leq k}$, respectively. Define $\bar{p} = \sum_{i=1}^{n} \nu_i p_i$. It is clear that $\bar{p}$ is the average of the fitted local models $\{p_i\}$. [66] observed that

$$\sum_{i=1}^{k} \nu_i K(p_i; p) = \sum_{i=1}^{k} \nu_i K(p_i; \bar{p}) + K(\bar{p}; p).$$

Hence, the minization problem (7.5) is equivalent to

$$p^* = \mathrm{argmin}_{p \in \mathcal{F}} K(\bar{p}; p). \tag{7.6}$$

If the distribution family $\mathcal{F}$ is large enough to include $\bar{p}$, then $p^* = \bar{p}$. However, $\bar{p}$ usually is not easily interpretable, because the local models may be over-parametrized to guarantee enough information is available for fitting the global model. Furthermore, $\bar{p}$ may not easily interpretable either. Hence, in practice, one prefers to restrict $\mathcal{F}$ to be a family of parametric models. For example, $\mathcal{F}$ can be the collection of Gaussian mixture distributions with finite components. Then, (7.6) becomes an optimization problem which can be solved numerically. When this requires intensive computation, an alternative using Monte Carlo simulation can be used. Instead of directly minizing (7.6), a sample $\{x_i\}_{1 \leq i \leq m}$ is randomly generated from $\bar{p}$. When $\bar{p}$ is complicated, Markov Chain Monte Carlo (MCMC) algorithms can be employed to generate the sample. Based on the sample, (7.6) can be further approximated by the following maximum likelihood approach,

$$p^* = \max_{p \in \mathcal{F}} \frac{1}{m} \sum_{j=1}^{m} \log(p(x_i)). \tag{7.7}$$

Note that the objective function above is the likelihood function. For clustering, the distributions in $\mathcal{F}$ are usually of mixture structures. Hence, the EM algorithm can be used for computing the maximum likelihood estimate p^*.

The security of the procedure described above is controlled by the privacy constraints imposed on the local models $\{p_i\}$. Hence, the data sources decide what kind of local models they want to release to a centralized analyst. Note that the private data are never disclosed.

The key requirement for the correctness of this approach is that the same underlying model generates the data at each source. In other words, the resources are homogeneous; differences between the local models are due to random variations rather than differences in the populations held by those sources. This assumption may be unrealistic in many applications, further work is needed to determine when this approach is applicable (or how to determine if it is applicable.)

7.2 Cryptography-based Approaches

The Secure Multiparty Computation model has also been used to develop descriptive modeling techniques.

7.2.1 EM-clustering for Horizontally Partitioned Data

Expectation Maximization (EM) is one of the most well-known clustering techniques, and can be viewed as a generalization of k-means clustering. The basic idea behind EM clustering is as follows. Assume the data to be clustered $\mathbf{y} = \{y_1, \cdots, y_n\}$ is independent and independently distributed, and drawn from a population with density function $f(\mathbf{y}; \Psi)$, where Ψ is a vector of the unknown parameters. The observed data log likelihood is:

$$\log L(\Psi) = \log f(\mathbf{y}; \Psi).$$

The maximum likelihood principle says that the estimators that maximize the data likelihood are consistent estimators of the true parameters. This means that we cluster by estimating the parameters Ψ, then clustering objects to their best-fit distribution among those parameters. Typically EM clustering assumes that data are drawn from a Gaussian distribution; the parameters Ψ to be learned are the mean and variance of each dimension.

It is generally infeasible to find analytical solutions to determine Ψ from the data. The EM algorithm is an iterative procedure to find the Ψ that maximizes

$\log L(\Psi)$ by data augmentation. The observed data $\mathbf{y}$ is augmented by the missing value $\mathbf{z}$ that contains group information of the observed data. More specifically, $\mathbf{z} = (Z_1, \cdots, Z_n)$ where $Z_j = (Z_{j1}, Z_{j2}, \cdots, Z_{jk})$. $Z_{ji} = 1$ means data point j belongs to the ith component. For instance, $Z_j = (1, 0, 0, 0, 0)$ means that the jth data point belongs to component 1. $\mathbf{x} = \langle \mathbf{y}, \mathbf{z} \rangle$ becomes complete data with density function $f_c(\mathbf{x}; \Psi)$. The complete data log likelihood is:

$$\log L_c(\Psi) = \log f_c(\mathbf{x}; \Psi).$$

Typically the complete data likelihood has a simpler structure and its expected likelihood can be maximized analytically.

Since we don't know the actual $\mathbf{z}$, the EM algorithm starts with a random assignment. It then iterates learning the parameters Ψ and updating the cluster assignments $\mathbf{z}$ until the cluster assignments stabilize. Dempster [17] proved that by maximizing $G(\Psi; \Psi^{(t)}) = E_{\Psi^{(t)}} \{\log L_c(\Psi)|\mathbf{y}\}$, the observed log likelihood is non-decreasing for each iteration step, which guarantees convergence of the algorithm. The algorithm contains two steps:

E-Step: On the $(t + 1)$st step, calculate the expected complete data log likelihood given observed data values: $G(\Psi; \Psi^{(t)})$
M-Step: Find $\Psi^{(t+1)}$ to maximize $G(\Psi; \Psi^{(t)})$.

Lin et al.[55] propose a privacy preserving EM algorithm for secure clustering of horizontally partitioned data. While this algorithm builds on many of the secure protocols we have already discussed, the fact that the EM algorithm is *iterative* makes the problem more challenging (as with the regression algorithm in Chapter 5.2.4.) A new twist is introduced in their proof that the intermediate steps, or simply the number of steps, does not reveal private information.

The key to this proof comes in two parts. First is showing that while the data disclosed at any iteration may exceed that needed for the solution, it does not reveal identifiable information. Second is showing that given two iterations, no private information is revealed that cannot be derived from only the second iteration.

We will now describe the algorithm and give an overview of the key novel features of the proof. To simplify the exposition, we consider only one dimension; i.e., data for only a single numerical attribute is collected by all parties.

Let the data y_j be partitioned across s sites ($1 \leq l \leq s$). Each site l has n_l data items.

To obtain a global estimation for the cluster i parameters at step $t + 1$ mean $\mu_i^{(t+1)}$, variance $\sigma_i^{2(t+1)}$, and proportion of items in i $\pi_i^{(t+1)}$ (the E step) requires only the global values n and

$$\mu_i^{(t+1)} = \sum_{j=1}^{n} z_{ij}^{(t)} y_j / \sum_{j=1}^{n} z_{ij}^{(t)} \tag{7.8}$$

$$\sigma_i^{(t+1)} = \sum_{j=1}^{n} z_{ij}^{(t)}(y_j - \mu_i^{(t+1)})^2 / \sum_{j=1}^{n} z_{ij}^{(t)} \qquad (7.9)$$

$$\pi_i^{(t+1)} = \sum_{j=1}^{n} z_{ij}^{(t)}/n \qquad (7.10)$$

Because of the commutativity of addition, we can rewrite the summations in the above equations as:

$$\sum_{j=1}^{n} z_{ij}^{(t)} y_j = \sum_{l=1}^{s} \sum_{j=1}^{n_l} z_{ijl}^{(t)} y_j \qquad (7.11)$$

$$\sum_{j=1}^{n} z_{ij}^{(t)} = \sum_{l=1}^{s} \sum_{j=1}^{n_l} z_{ijl}^{(t)} \qquad (7.12)$$

$$\sum_{j=1}^{n} z_{ij}^{(t)}(y_j - \mu_i^{(t+1)})^2 = \sum_{l=1}^{s} \sum_{j=1}^{n_l} z_{ijl}^{(t)}(y_j - \mu_i^{(t+1)})^2 \qquad (7.13)$$

Observe that the second summation in each of the above equations can be computed locally at each site:

$$A_{il} = \sum_{j=1}^{n_l} z_{ijl}^{(t)} y_j \qquad (7.14)$$

$$B_{il} = \sum_{j=1}^{n_l} z_{ijl}^{(t)} \qquad (7.15)$$

$$C_{il} = \sum_{j=1}^{n_l} z_{ijl}^{(t)}(y_j - \mu_i^{(t+1)})^2 \qquad (7.16)$$

Calculating local values does not disclose private values; secure summation (Chapter 3.3.2) shows how to compute the global summations without revealing anything except the final sums in Equations 7.11–7.13. Given these sums, and assuming n is known to all (or computed securely using a secure summation), each site can calculate μ_i, σ_i, and π_i locally using Equations 7.8–7.10.

Given the parameters for each cluster, each site can locally perform the next E-step. This is done by assigning its local items to clusters as follows:

$$z_{ijl} = \frac{\pi_i^{(t+1)} f_i(y_{jl}; \mu_i^{(t+1)}, \sigma_i^{(t+1)})}{\sum_i \pi_i^{(t+1)} f_i(y_{jl}; \mu_i^{(t+1)}, \sigma_i^{(t+1)})} \qquad (7.17)$$

where y_{jl} is a data point at site l.

The E-step and M-step iterate until

$$|L^{(t+1)} - L^{(t)}| \le \epsilon. \qquad (7.18)$$

where

$$L^{(t)}(\theta^{(t)}, \mathbf{z}^{(t)}|y) = \sum_{j=1}^{n}\sum_{i=1}^{k}\{z_{ij}^{(t)}[\log \pi_i f_i(y_j^{(t)}|\theta^{(t)})]\}. \qquad (7.19)$$

This can again be calculated using secure summation and a secure comparison, as with the association rule algorithm in Chapter 6.2.1. Algorithm 2 summarizes the method.

Algorithm 2 Secure EM Algorithm.

At each site l, $\forall_{i=1..n_l,j=1..k}$ randomly initialize z_{ijl} to 0 or 1.
Use secure sum of Section 3.3.2 to compute $n = \sum_{l=1}^{s} n_l$
$t = 0$
while Threshold criterion of log likelihood not met **do**
    for all $i = 1..k$ **do**
        At each site l, calculate $A_{il}^{(t+1)}$ and $B_{il}^{(t+1)}$ using equations (7.14) and (7.15).
        Use secure sum to calculate $A_i^{(t+1)}$ and $B_i^{(t+1)}$.
        Site 1 uses these to compute $\mu_i^{(t+1)}$ and broadcasts it to all sites.
        Each site l calculates $C_{il}^{(t+1)}$ using equation (7.16).
        Use secure sum to calculate $C_i^{(t+1)}$.
        Site 1 calculates $\sigma_i^{2(t+1)}$ and $\pi_i^{(t+1)}$ and broadcasts them to all sites.
        At each site l, $\forall_{j=1..n_l}$ update $z_{ijl}^{(t+1)}$ using equation (7.17).
    end for
    $t = t + 1$
    Calculate the log likelihood difference using equation (7.18) and (7.19).
end while

Secure summation guarantees that the only information disclosed at each iteration are the parameters n, μ_i, σ_i, and π_i. (Technically, the sums used to calculate those parameters are disclosed, but note that the sums can be derived from the final parameter values.) The first question is the privacy implications of revealing these global parameters. In [55] it is shown that these values do not directly reveal individually identifiable information, or even site identifiable information. Even in exceptional cases, such as a cluster containing a single individual (thus leading the cluster center to converge to the values for that individual), only the site containing that individual knows which cluster the individual is in. Since that site already knows the data for that individual, there is no privacy breach. Similar arguments are given for the other parameters.

A more challenging problem is posed by the ability to compare values across iterations. While a single mean may not reveal anything, does the change in that mean across iterations reveal too much about the data? To address this question, assume without loss of generality that s new data points are assigned to component i. From the mean and variance of steps t and $t+1$, we have:

$$\sigma_i^{(t)} = \frac{\sum_{j=1}^{n_i}(y_j - \mu_i^{(t)})^2}{n_i - 1}$$

$$\sigma_i^{(t+1)} = \frac{\sum_{j=1}^{n_i}(y_j - \mu_i^{(t+1)})^2 + \sum_{j=n_i+1}^{n_i+s}(y_j - \mu_i^{(t+1)})(y_j - \mu_i^{(t+1)})'}{n_i + s - 1}$$

When $s > 1$, these two equations have infinite solutions for $y_{n_i+1}, \ldots, y_{n_i+s}$. In other words, values from previous iterations will not give specific information on individuals that is not already available from the final iteration.

7.2.2 K-means Clustering for Vertically Partitioned Data

Vertically partitioned data once again raises its own complications. First, we must decide how the model is shared by the parties. It is easy to say that we group the data points into clusters. However, apart from cluster membership, what other information is shared by the parties? Do all parties know all the relevant information about each cluster or do they know only the information restricted to their attributes? Which of these alternatives makes sense in the particular real-life situation? These issues are common to any privacy-preserving clustering over vertically partitioned data.

Vaidya and Clifton [81] proposed the first method for clustering over vertically partitioned data – a privacy-preserving protocol perform do k-means clustering. Though all parties know the final assignment of data points to clusters, they retain only partial information for each cluster. The cluster centers μ_i are assumed to be semiprivate information, i.e., each site can learn only the components of μ that correspond to the attributes it holds. Thus, all information about a site's attributes (not just individual values) is kept private; if sharing the μ is desired, an evaluation of privacy/secrecy concerns can be performed after the values are known.

K-means clustering[25, 34] is a simple technique to group items into k clusters. The basic idea behind k-means clustering is as follows:

1: Initialize the k means $\mu_1 \ldots \mu_k$ to 0.
2: Arbitrarily select k starting points $\mu'_1 \ldots \mu'_k$
3: **repeat**
4:     Assign $\mu'_1 \ldots \mu'_k$ to $\mu_1 \ldots \mu_k$ respectively
5:     **for all** points i **do**
6:         Assign point i to cluster j if distance $d(i, \mu_j)$ is the minimum over all j.
7:     **end for**
8:     Calculate new means $\mu'_1 \ldots \mu'_k$.
9: **until** the difference between $\mu_1 \ldots \mu_k$ and $\mu'_1 \ldots \mu'_k$ is acceptably low.

Each item is placed in its closest cluster, and the cluster centers are then adjusted based on the data placement. This repeats until the positions stabilize.

Let us now see what is necessary to make this algorithm privacy-preserving. Since k is traditionally used to denote the number of clusters, in the following discussion, we abuse notation a bit, and just for this section, use r to refer to the number of parties. Step $1, 2$ are quite simple. Since the data is vertically partitioned, each party can randomly choose its part for each of the k means. The assignments in step 4 are local assignments so they can be easily carried out by each party. If all of the points have been appropriately assigned to the closest cluster, calculating the new means is again a local operation. Thus, each party can compute the new mean for its attributes for each cluster. Figuring out when to stop (Step 9 is the other step requiring work. Thus, in order to make the entire algorithm privacy-preserving, the following two computations need to be carried out securely:

- The closest cluster computation for each point
- The termination test: is the improvement to the mean approximation in the iteration below a threshold?

Closest cluster computation

The closest cluster computation is invoked for every single data point in each iteration. Each party has as its input the component of the distance corresponding to each of the k clusters. This is equivalent to having a matrix of distances of dimension $k \times r$. For common distance metrics; such as Euclidean, Manhattan, or Minkowski; this translates to finding the cluster where the sum of the local distances is the minimum among all the clusters.

The problem is formally defined as follows. Consider r parties $P_1, \ldots, P_r$, each with their own k-element vector $\mathbf{X_i}$:

$$P_1 \text{ has } \mathbf{X_1} = \begin{bmatrix} x_{11} \\ x_{21} \\ \vdots \\ x_{k1} \end{bmatrix}, P_2 \text{ has } \begin{bmatrix} x_{12} \\ x_{22} \\ \vdots \\ x_{k2} \end{bmatrix}, \ldots, P_r \text{ has } \begin{bmatrix} x_{1r} \\ x_{2r} \\ \vdots \\ x_{kr} \end{bmatrix}.$$

The goal is to compute the index l that represents the row with the minimum sum. Formally, find

$$argmin_{i=1..k} \left(\sum_{j=1..r} x_{ij} \right)$$

For use in k-means clustering, $x_{ij} = |\mu_{ij} - point_j|$, or site P_j's component of the distance between a point and the cluster i with mean μ_i.

The security of the algorithm is based on three key ideas.

1. Disguise the site components of the distance with random values that cancel out when combined.
2. Compare distances so only the comparison result is learned; no party knows the distances being compared.
3. Permute the order of clusters so the real meaning of the comparison results is unknown.

The algorithm also requires three non-colluding sites. These parties may be among the parties holding data, but could be external as well. They need only know the number of sites r and the number of clusters k. Assuming they do not collude with each other, they learn nothing from the algorithm. For simplicity of presentation, we will assume the non-colluding sites are P_1, P_2, and P_r among the data holders. Using external sites, instead of participating sites P_1, P_2 and P_r, to be the non-colluding sites, is trivial.

The algorithm proceeds as follows. Site P_1 generates a length k random vector $\mathbf{V_i}$ for each site i, such that $\sum_{i=1}^r \mathbf{V_i} = \mathbf{0}$. P_1 also chooses a permutation π of $1..k$. P_1 then engages each site P_i in the permutation algorithm to generate the sum of the vector $\mathbf{V_i}$ and P_i's distances $\mathbf{X_i}$. The resulting vector is known only to P_i, and is permuted by π known only to P_1, i.e., P_i has $\pi(\mathbf{V_i} + \mathbf{X_i})$, but does not know π or $\mathbf{V_i}$. P_i and $P_3 \ldots P_{r-1}$ send their vectors to P_r.

Sites P_2 and P_r now engage in a series of secure addition / comparisons to find the (permuted) index of the minimum distance. Specifically, they want to find if $\sum_{i=1}^r x_{li} + v_{li} < \sum_{i=1}^r x_{mi} + v_{mi}$. Since $\forall l, \sum_{i=1}^r v_{li} = 0$, the result is $\sum_{i=1}^r x_{li} < \sum_{i=1}^r x_{mi}$, showing which cluster (l or m) is closest to the point. P_r has all components of the sum except $\mathbf{X_2} + \mathbf{V_2}$. For each comparison, we use a secure circuit evaluation (see Chapter 3.3.1) that calculates $a_2 + a_r < b_2 + b_r$, without disclosing anything but the comparison result. After $k - 1$ such comparisons, keeping the minimum each time, the minimum cluster is known.

P_2 and P_r now know the minimum cluster in the permutation π. They do not know the real cluster it corresponds to (or the cluster that corresponds to any of the others items in the comparisons.) For this, they send the minimum i back to site P_1. P_1 broadcasts the result $\pi^{-1}(i)$, the proper cluster for the point.

The permutation algorithm is one of the two key building blocks borrowed from the Secure Multiparty Computation literature. The secure permutation algorithm developed by Du and Atallah[20] simultaneously computes a vector sum and permutes the order of the elements in the vector. The key behind the solution is the use of Homomorphic encryption. An encryption function $\mathcal{H} : \mathcal{R} \rightarrow \mathcal{S}$ is called *additively homomorphic* if there is an efficient algorithm *Plus* to compute $H(x + y)$ from $H(x)$ and $H(y)$ that does not reveal x or y. Many such systems exist; examples include systems by Benaloh[10], Naccache and Stern [61], Okamoto and Uchiyama[65], and Paillier [68]. This allows us to perform addition of encrypted data without decrypting it. Basically, the first party encrypts its data using homomorphic encryption and sends to the other party the encrypted data as well as the encryption key. Using the encryption key, the other party can encrypt random numbers. Using the homomorphic property of the encryption, these two encrypted numbers can be added. Now the other party permutes the resulting sum vector and sends it back to the original party which can decrypt to get the permuted vector with randoms added. A graphical depiction of stages 1 and 2 is given in Figures 7.4 and 7.5. More details along with a security proof can be found in [81].

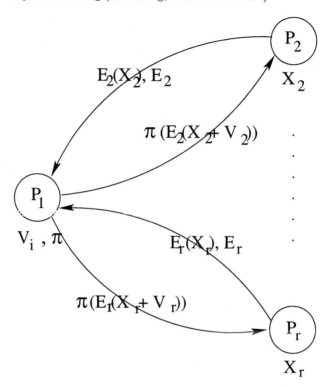

Fig. 7.4. Closest Cluster - Stage 1

The other primitive used from SMC is the secure addition comparison, which builds a circuit that has two inputs from each party, sums the first input of both parties and the second input of both parties, and returns the result of comparing the two sums. This (simple) circuit is evaluated securely using the generic algorithm described in Section 3.3.1.

The termination test

When to terminate is decided by comparing the improvement to the mean approximation in each iteration to a threshold. If the improvement is sufficient, the algorithm proceeds, otherwise it terminates.

Each party locally computes the difference between its share of the old mean and the new mean for each of the k clusters. Now, the parties must figure out if the total sum is less than the threshold. This looks straightforward, except that to maintain security (and practicality) all arithmetic takes place in a field and is thus modular arithmetic. This results in a non-obvious threshold evaluation at the end, consisting of a secure addition/comparison. *Intervals* are compared rather than the actual numbers. Further details can be found in [81].

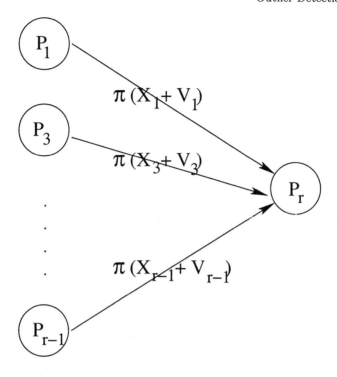

Fig. 7.5. Closest Cluster - Stage 2

Even now, this protocol is not fully secure since some intermediate results are leaked. Essentially, the intermediate cluster assignment of data points is known to every party for each iteration, though the final result only specifies the *final* clusters. However, this compromise is required for efficiency, though finding better solution is an issue for future research.

7.3 Outlier Detection

Outlier detection has wide application; one that has received considerable attention is the search for terrorism. Detecting previously unknown suspicious behavior is a clear outlier detection problem. The search for terrorism has also been the flash point for attacks on data mining by privacy advocates; the U.S. Terrorism Information Awareness program was killed for this reason[54].

Outlier detection has numerous other applications that also raise privacy concerns. Mining for anomalies has been used for network intrusion detection[8, 53]; privacy advocates have responded with research to enhance anonymity[74, 38]. Fraud discovery in the mobile phone industry has also made use of outlier detection[29]; organizations must be careful to avoid overstepping the bounds of privacy legislation[26]. Privacy-preserving outlier detection will ensure these concerns are balanced, allowing us to get the benefits of outlier detection without being thwarted by legal or technical counter-measures. However, outlier detection, by definition, means pinpointing entities/transactions that are anomalous. In no sense does it summarize information. Thus, the entities that are highlighted lose all privacy at the individual level. This is necessary for the true positives in any case. The problem remains for the false positives – entities identified as outliers without really being so. This seems problematic, however, a couple of caveats exist. First, any detection technique is not fool-proof and false positives always exist. We merely reduce the privacy leakage and problems. Secondly, technical solutions exist. All the identifiers can be eliminated to begin with. The outliers detected are hand examined and if sufficient cause exists, the anonymization is taken away and the real identity is revealed (just as it occurs in real life with a court order).

An obvious question again is to ask why privacy-preserving outlier detection is necessary in the first place? Why not run the outlier detection algorithm locally at each site and then combine the results? This is especially true since false negatives will never occur due to the local detection process as long as a fixed threshold is used. For example, a person who is close to 5 points in a local dataset, will be close to at least 5 points in the global dataset as well. As long as the outlier detection threshold stays at 4, this point will never be detected as an outlier. This does make sense in some applications. However, the problem of false positives does remain. Figure 7.6 demonstrates this problem. Here we have horizontal partitioning of data between two parties (diamonds and rectangles are used to show the different points). If outlier detection was carried out locally at diamond's site, among others, the two points circled would be detected as outliers. In the global case, it can be clearly seen that the point at $(0.1, 0.9)$ is truly an outlier where as the point at $(0.81, 0.85)$ is not an outlier. By doing outlier detection on the global dataset, detecting this false positive would be avoided. This is especially important for applications such as terrorist detection – where the points are many, and the resources to manually check them later, small.

As usual, the basic assumption is that data is distributed; the stewards of the data are allowed to use it, but disclosing it to others is a privacy violation. While there are numerous different definitions of outliers as well as techniques to find them, the only one currently developed in a privacy-preserving fashion is for distance-based outliers. The method developed by Vaidya and Clifton[83] finds distance-based outliers without any party gaining knowledge beyond learning which items are outliers. Ensuring that data is not disclosed maintains privacy, i.e., no privacy is lost beyond that inherently revealed in knowing the outliers. Even knowing which items are outliers need not be revealed to all parties, further preventing privacy breaches.

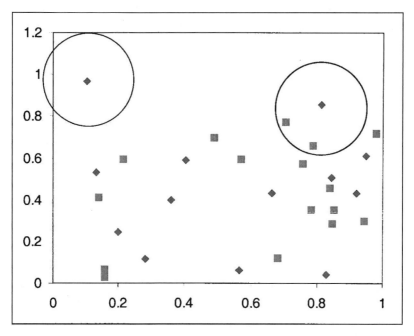

Fig. 7.6. Global outlier detection on data set collected from two parties

7.3.1 Distance-based Outliers

Knorr and Ng [51] define the notion of a Distance Based outlier as follows: *An object O in a dataset T is a DB(p,dt)-outlier if at least fraction p of the objects in T lie at distance greater than dt from O.* Other distance based outlier techniques also exist[52, 73]. The advantages of distance based outliers are that no explicit distribution needs to be defined to determine unusualness, and that it can be applied to any feature space for which we can define a distance measure. Euclidean distance is the standard, although the algorithms are easily extended to general Minkowski distances. There are other non distance based techniques for finding outliers as well as significant work in statistics [9], but there is no work on finding them in a privacy-preserving fashion – thus, this is a rich area for future work.

For Euclidean distance, for vertically partitioned data, the distance dt is fixed by the local parties deciding on the local distances dt_i (i.e., $dt = \sum_{i=1}^{k} dt_i$), since no site globally knows all of the attributes. An object X is an outlier if at least $p\%$ of the other objects lie at a distance greater than dt. For horizontally partitioned data, all parties together decide on the distance threshold dt. Rather than computing the distance and comparing against the threshold, the difference between the distance squared and the threshold squared is computed. This difference is calculated, and the number of objects of each party being to which this object is relatively an outlier is computed.

Now all parties simply have to sum these values up and if the sum is greater than $p\%$, then that object is an outlier.

7.3.2 Basic Approach

The approach duplicates the results of the outlier detection algorithm of [51]. The idea is that an object O is an outlier if more than a percentage p of the objects in the data set are farther than distance dt from O. The basic idea is that parties compute the portion of the answer they know, then engage in a secure sum to compute the total distance. The key is that this total is (randomly) split between sites, so nobody knows the actual distance. A secure protocol is used to determine if the actual distance between any two points exceeds the threshold; again the comparison results are randomly split such that summing the splits (over a closed field) results in a 1 if the distance exceeds the threshold, or a 0 otherwise.

For a given object O, each site can now sum all of its shares of comparison results (again over the closed field). When added to the sum of shares from other sites, the result is the correct count; all that remains is to compare it with the percentage threshold p. This addition/comparison is also done with a secure protocol, revealing only the result: if O is an outlier.

The pairwise comparison of all points may seem excessive, but early termination could disclose information about relative positions of points (this will be discussed further in Section 7.3.7.) The asymptotic complexity still equals that of [51].

Note that a secure solution requires that all operations are carried out modulo some field. For the algorithms, the field D is used for distances, and F is used for counts of the number of entities. The field F must be over twice the number of objects. Limits on D are based on maximum distances; details on the size are given with each algorithm.

The following Sections presents the privacy-preserving algorithms for both horizontally and vertically partitioned data. Following that, the complete proof of security for these algorithms is presented. This is especially instructive for readers wishing to develop their own algorithms since the proof of security forms a significantly important component necessary for trust in the overall solution. A discussion of the computational and communication complexity of the algorithms rounds off this section, and affords the opportunity to discuss avenues for future work in this area.

7.3.3 Horizontally Partitioned Data

The key idea behind the algorithm for horizontally partitioned data is as follows. For each object i, the protocol iterates over every other object j. If the same party holds both i and j, it can easily find the distance and compare against the threshold. If two different parties hold the two objects, the parties engage in a distance calculation protocol (Section 7.3.3) to get *random shares* of the distance. Now, a second protocol compares the shares with the threshold, returning 1 if the distance exceeds the threshold, or 0 if it does not. The key to this second protocol is that the 1 or 0 is actually two shares r'_q and r'_s returned to the two parties, such that $r'_q + r'_s = 1$ (or 0) (mod F). Looking at only one share, neither party can learn anything.

Once all points have been compared, the parties individually sum their shares. Since the shares add to 1 for distances exceeding the distance threshold, and 0 otherwise, the total sum   (mod F) gives the number of points for which the distance exceeds the threshold. Explicit computation of this sum would still reveal the actual number of points distant. So the parties do not actually compute this sum; instead all parties pass their (random) shares to a designate to add, and the designated party and the party holding the point engage in a secure protocol that reveals only if the sum of the shares exceeds $p\%$. Thus, the only result of the protocol is to reveal whether the point is an outlier or not.

Algorithm 3 gives the complete details. Steps 3-15 are the pairwise comparison of two points, giving each party random shares of a 1 (if the points are far apart) or 0 (if the points are within the distance threshold dt). The random split of shares ensures that nothing is learned by either party. In steps 16-18, a party P_v other than P_q (i.e., any party other than the party holding the object being evaluated) is chosen. All other parties (other than P_q and P_v) send their shares to P_v who sums these with its share. Again, since each share is a random split (and P_q holds the other part of the split), no party learns anything. Finally, P_v and P_q add and compare their shares, revealing only if the object o_i is an outlier. Note that the shares of this comparison are split, and could be sent to any party (P_q in Algorithm 3, but it need not even be one of the P_r parties). Only that party (e.g., a fraud prevention unit) learns if o_i is an outlier, the others learn nothing.

Computing distance between two points

A key step of Algorithm 3 requires computation of the distance between two objects. However, rather than simply revealing the result to the two parties the computation must output random shares of that distance to the two parties. This distance is than compared against the threshold to determine whether the two objects are close or not. For convenience, what is actually computed are shares of the *square* of the distance. These are then compared with the square of the threshold. (This does not change the result, since squaring is a monotonically increasing function.) We now look at the algorithm for computing shares of the square of the Euclidean distance. The algorithm is based on a secure scalar product, for which many protocols exist [20, 80, 43, 35]; any of these can be used.

Formally, assume that there are two parties, P_1 and P_2. All computations are over a field D larger than the square of the maximum distance. P_1's input is the point X, P_2's input is the point Y. The outputs are r_1 and r_2

Algorithm 3 Finding DB(p,D)-outliers

Require: k parties, $P_1, \ldots, P_k$; each holding a subset of the objects O.
Require: Fields D larger than the maximum distance squared, F larger than $|O|$
 1: **for all** objects $o_i \in O$ {Let P_q be the party holding o_i} **do**
 2:     Every party P_r initializes local counter num_r to 0
 3:     **for all** objects $o_j \in O$, $o_j \neq o_i$ **do**
 4:         **if** P_q holds o_j **then**
 5:             **if** $Distance(o_i, o_j) > dt$ {Computed locally at P_q} **then**
 6:                 P_q locally increments num_q (modulo F)
 7:             **end if**
 8:         **else**
 9:             {Assume party P_s holds o_j}
10:             P_q and P_s engage in the distance computation protocol (Section 7.3.3) to get r_q and r_s respectively such that $r_q + r_s \pmod{D} = Distance^2(o_i, o_j)$
11:             P_q and P_s use the secure comparison protocol (Section 7.3.5) to get r'_q and r'_s respectively such that the following condition holds: if $r_q + r_s \pmod{D} > dt^2$, then $r'_q + r'_s = 1 \pmod{F}$, else $r'_q + r'_s = 0 \pmod{F}$.
12:         **end if**
13:         At P_q: $num_q \leftarrow num_q + r'_q$
14:         At P_s: $num_s \leftarrow num_s + r'_s$
15:     **end for**
16:     {Let P_v be a party other than P_q}
17:     All parties P_r excepting P_q and P_v send their local counters num_r to P_v
18:     At P_v: $num_v \leftarrow \sum_{i \neq q} num_i$
19:     P_q and P_v use the secure comparison protocol (Section 7.3.5) to get $temp_q$ and $temp_v$ respectively such that the following condition holds: if $num_q + num_v \pmod{F} > |O| * p\%$, then $temp_q + temp_v = 1$ (o_i is an outlier), otherwise $temp_q + temp_v = 0$
20:     P_v sends $temp_v$ to P_q, revealing to P_q if o_i is an outlier.
21: **end for**

---

respectively (independently uniformly distributed over D), such that $r_1 + r_2 = Distance^2(X, Y) \pmod{D}$, where $Distance(X, Y)$ is the Euclidean distance between the points X and Y.

Let there be m attributes, and a point W be represented by its m-dimensional tuple $(w_1, \ldots, w_m)$. Each co-ordinate represents the value of the point for that attribute. The square of the Euclidean distance between X and Y is given by

$$
\begin{aligned}
Distance^2(X, Y) &= \sum_{r=1}^{m} (x_r - y_r)^2 \\
&= x_1^2 - 2x_1 y_1 + y_1^2 + \ldots \\
&\quad \ldots + x_m^2 - 2x_m y_m + y_m^2 \\
&= \sum_{r=1}^{m} x_r^2 + \sum_{r=1}^{m} y_r^2 - \sum_{r=1}^{m} 2x_r y_r
\end{aligned}
$$

P_1 can independently calculate $\sum_r x_r^2$. Similarly, P_2 can calculate $\sum_r y_r^2$. As long as there is more than one attribute (i.e., $m > 1$), the remaining sum $\sum_r (2x_r)(-y_r)$ is simply the scalar product of two m-dimensional vectors. P_1 and P_2 engage in a secure scalar product protocol to get random shares of the dot product. This, added to their prior calculated values, gives each party a random share of the square of the distance.

Assuming that the scalar product protocol is secure, applying the composition theorem of [36] shows that the entire protocol is secure.

7.3.4 Vertically Partitioned Data

Vertically partitioned data introduces a different challenge. Since each party owns some of the attributes, each party can compute the distance between two objects for those attributes. Thus, each party can compute a *share* of the pairwise distance locally; the sum of these shares is the total distance. However, revealing the distance still reveals too much information, therefore a secure protocol is used to get shares of the pairwise comparison of distance and threshold. From this point, it is similar to horizontal partitioning: Add the shares and determine if they exceed $p\%$.

An interesting side effect of this algorithm is that the parties need not reveal any information about the attributes they hold, or even the number of attributes. Each party locally determines the distance threshold for its attributes (or more precisely, the share of the overall threshold for its attributes). Instead of computing the local pairwise distance, each party computes the difference between the local pairwise distance and the local threshold. If the sum of these differences is greater than 0, the pairwise distance exceeds the threshold.

Algorithm 4 gives the full details. In steps 6-10, the sites sum their local distances (actually the difference between the local distance and the local threshold). The random x added by P_1 masks the distance from each party. In steps 11-13, Parties P_1 and P_k get shares of the pairwise comparison result, as in Algorithm 3. The comparison is a test if the sum is greater than 0 (since the threshold has already been subtracted.) These two parties keep a running sum of their shares. At the end, in step 15 these shares are added and compared with the percentage threshold, again as in Algorithm 3.

Theorem 7.1. *Proof of Correctness: Algorithm 4 correctly returns as output the complete set of points that are global outliers.*

Proof. In order to prove the correctness of Algorithm 4, it is sufficient to prove that a point is reported as an outlier if and only if it is truly an outlier. Consider point q. If q is an outlier, in step 11 for at least $p\% * |O| + 1$ of the other points, $m_1 + m_k = 1 \pmod{F}$. Since $|F| > |O|$, it follows that $m_1' + m_k' > |O| * p\%$. Therefore, point q will be correctly reported as an outlier. If q is not an outlier, the same argument applies in reverse. Thus, in

---

Algorithm 4 Finding DB(p,D)-outliers

Require: k parties, $P_1, \ldots, P_k$; each holding a subset of the attributes for all objects O.

Require: dt_r : local distance threshold for P_r (e.g., $dt^2 + m_r/m$).

Require: Fields D larger than twice the maximum distance value (e.g., for Euclidean this is actually $Distance^2$), F larger than $|O|$

1: **for all** objects $o_i \in O$ **do**

2:    $m'_1 \leftarrow m'_k \leftarrow 0 \pmod{F}$

3:    **for all** objects $o_j \in O, o_j \neq o_i$ **do**

4:       P_1: Randomly choose a number x from a uniform distribution over the field D

5:       P_1: $x' \leftarrow x$

6:       **for** $r \leftarrow 1, \ldots, k-1$ **do**

7:          At P_r: $x' \leftarrow x' + Distance_r(o_i, o_j) - dt_r \pmod{D}$ {$Distance_r$ is local distance at P_r}

8:          P_r sends x' to P_{r+1}

9:       **end for**

10:      At P_k: $x' \leftarrow x' + Distance_k(o_i, o_j) - dt_k \pmod{D}$

11:      P_1 and P_k engage in the secure comparison protocol (Section 7.3.5) to get m_1 and m_k respectively such that the following condition holds: if $0 < x' + (-x) \pmod{D} < |D|/2$, then $m_1 + m_k = 1 \pmod{F}$, otherwise $m_1 + m_k = 0 \pmod{F}$

12:      At P_1: $m'_1 \leftarrow m'_1 + m_1 \pmod{F}$

13:      At P_k: $m'_k \leftarrow m'_k + m_k \pmod{F}$

14:   **end for**

15:   P_1 and P_k engage in the secure comparison protocol (Section 7.3.5) to get $temp_1$ and $temp_k$ respectively such that the following condition holds: if $m'_1 + m'_k \pmod{F} > |O|*p\%$, then $temp_1 + temp_k \leftarrow 1$ (o_i is an outlier), otherwise $temp_1 + temp_k \leftarrow 0$

16:   P_1 and P_k send $temp_1$ and $temp_k$ to the party authorized to learn the result; if $temp_1 + temp_k = 1$ then o_i is an outlier.

17: **end for**

---

step 11 at most $p\% * |O| - 1$ points, $m_1 + m_k = 1 \pmod{F}$. Again, since $|F| > |O|$, it follows that $m'_1 + m'_k \leq |O| * p\%$. Therefore, point q will not be reported as an outlier.

7.3.5 Modified Secure Comparison Protocol

At several stages in the algorithm, a protocol is required to securely compare the sum of two numbers, with the output split between the parties holding those numbers. This can be accomplished using the generic circuit evaluation technique first proposed by Yao[90]. Formally, we need a modified secure comparison protocol for two parties, A and B. The local inputs are x_a and x_b and the local outputs are y_a and y_b. All operations on input are in a field F_1 and output are in a field F_2. $y_a + y_b = 1 \pmod{F_2}$ if $x_a + x_b \pmod{F_1} > 0$, otherwise $y_a + y_b = 0 \pmod{F_2}$. A final requirement is that y_a and y_b should be independently uniformly distributed over F (clearly the joint distribution is not uniform).

The standard secure multiparty computation circuit-based approach[36] can be used to solve this problem. Effectively, A chooses y_a with a uniform distribution over F, and provides it as an additional input to the circuit that appropriately computes y_b. The circuit is then securely evaluated, with B receiving the output y_b. The complexity of this is equivalent to the complexity of Yao's Millionaire's problem (simple secure comparison).

7.3.6 Security Analysis

The proof technique used is that of Secure Multiparty Computation (Section 3.3). The idea is that since what a party sees during the protocol (its shares) are randomly chosen from a uniform distribution over a field, it learns nothing in isolation. (Of course, collusion with other parties could reveal information, since the *joint* distribution of the shares is not random). The idea of the proof is based on a simulation argument: If we can define a simulator that uses the algorithm output and a party's own data to simulate the messages seen by a party during a real execution of the protocol, then the real execution isn't giving away any new information (as long as the simulator runs in polynomial time).

The formal definitions for this can be found in [36] and are discussed in Section 3.3. We now look at the proof of security for Algorithms 3 and 4 under this framework.

Horizontally partitioned data

Theorem 7.2. *Algorithm 3 returns as output the set of points that are global outliers, and reveals no other information to any party provided parties do not collude.*

Proof. Presuming that the number of objects $|O|$ is known globally, each party can locally set up and run its own components of Algorithm 3 (e.g., a party only needs to worry about its local objects in the "For all objects" statements at lines 1 and 3.) In the absence of some type of secure anonymous send[74, 38] (e.g., anonymous transmission with public key cryptography to ensure reception only by the correct party), the number of objects at each site is revealed. Since at least an upper bound on the number of items is inherently revealed by the running time of the algorithm, we assume these values are known.

The next problem is to simulate the messages seen by each party during the algorithm. Communication occurs only at steps 10, 11, 17, 19, and 20. Now we look at how each step can be simulated.

Step 10:

P_q and P_s each receive a share of the square of the distance. As can be seen in Section 7.3.3, all parts of the shares are computed locally except for shares of the scalar product. Assume that the scalar product protocol chooses shares by selecting the share for P_q (call it s_q) randomly from a uniform distribution over D. Then $\forall x \in D, Pr(s_q = x) = \frac{1}{|D|}$. Thus, s_q is easily simulated by simply choosing a random value from D. Let the result $w = \sum_r (2x_r)(-y_r)$ be fixed. Then $\forall x \in D, Pr(s_s = x) = Pr(w - s_q = y) = Pr(s_q = w - y) = \frac{1}{|D|}$. Therefore, the simulator for P_s can simulate this message by simply choosing a random number from an uniform distribution over D. Assuming that the scalar product protocol is secure, applying the composition theorem shows that step 10 is secure.

Steps 11 and 19:

The simulator for party P_q (respectively P_s) chooses a number randomly from a uniform distribution, this time over the field F. By the same argument as above, the actual values are uniformly distributed, so the probability of the simulator and the real protocol choosing any particular value are the same. Since a circuit for secure comparison is used (and all parameters – $dt, p, |O|$ are known to all parties) , using the composition theorem, no additional information is leaked and step 11 (respectively 19) is secure.

Step 17:

P_v receives several shares num_r. However, note that num_r is a sum, where all components of the sum are random shares from Step 11. Since P_v receives only shares from the P_s in step 11, and receives none from P_q, all of the shares in the sum are independent. As long as P_q does not collude with P_v, the independence assumption holds. Thus the sum num_r can be simulated by choosing a random value from a uniform distribution over F.

Step 20:

Since P_q knows the final result (1 if o_i is an outlier, 0 otherwise), and $temp_q$ was simulated in step 19, it can simulate $temp_v$ with the results (1 or 0) $-temp_q \bmod F$.

The simulator clearly runs in polynomial time (the same as the algorithm). Since each party is able to simulate the view of its execution (i.e., the probability of any particular value is the same as in a real execution with the same inputs/results) in polynomial time, the algorithm is secure with respect to Definition 3.1.

While the proof is formally only for the semi-honest model, it can be seen that a malicious party in isolation cannot learn private values (regardless of what it does, it is still possible to simulate what it sees without knowing the input of the other parties.) This assumes that the underlying scalar product and secure comparison protocols are secure against malicious behavior. A malicious party can cause incorrect results, but it cannot learn private data values.

Vertically partitioned data

Theorem 7.3. *Algorithm 4 returns as output the set of points that are global outliers while revealing no other information to any party, provided parties do not collude.*

Proof. All parties know the number (and identity) of objects in O. Thus they can set up the loops; the simulator just runs the algorithm to generate most of the simulation. The only communication is at lines 8, 11, 15, and 16.

Step 8:

Each party P_s sees $x' = x + \sum_{r=1}^{s-1} Distance_r(o_i, o_j)$, where x is the random value chosen by P_1. $Pr(x' = y) = Pr(x + \sum_{r=1}^{s-1} Distance_r(o_i, o_j) = y) = Pr(x = y - \sum_{r=0}^{s-1} Distance_r(o_i, o_j)) = \frac{1}{|D|}$. Thus we can simulate the value received by choosing a random value from a uniform distribution over D.

Steps 11 and 15:

Each step is again a secure comparison, so messages are simulated as in Steps 11 and 19 of Theorem 7.2.

Step 16:

This is again the final result, simulated as in Step 20 of Theorem 7.2. $temp_1$ is simulated by choosing a random value, $temp_k = result - temp_1$. By the same argument on random shares used above, the distribution of simulated values is indistinguishable from the distribution of the shares.

Again, the simulator clearly runs in polynomial time (the same as the algorithm). Since each party is able to simulate the view of its execution (i.e., the probability of any particular value is the same as in a real execution with the same inputs/results) in polynomial time, the algorithm is secure with respect to Definition 3.1.

Without collusion and assuming a malicious-model secure comparison, a malicious party is unable to learn anything it could not learn from altering its input. Step 8 is particularly sensitive to collusion, but can be improved (at cost) by splitting the sum into shares and performing several such sums (see [47] for more discussion of collusion-resistant secure sum).

7.3.7 Computation and Communication Analysis

In general we do not discuss the computational/communicational complexity of the algorithms in detail. However, in this case the algorithmic complexity raises interesting issues vis-a-vis security. Therefore we discuss it below in detail.

Both Algorithms 3 and 4 suffer the drawback of having quadratic computation complexity due to the nested iteration over all objects.

Due to the nested iteration, Algorithm 3 requires $O(n^2)$ distance computations and secure comparisons (steps 10-11), where n is the total number of objects. Similarly, Algorithm 4 also requires $O(n^2)$ secure comparisons (step 11). While operation parallelism can be used to reduce the round complexity of communication, the key practical issue is the computational complexity of the encryption required for the secure comparison and scalar product protocols.

This quadratic complexity is troubling since the major focus of new algorithms for outlier detection has been to reduce the complexity, since n^2 is assumed to be inordinately large. However, achieving lower than quadratic complexity is challenging – at least with the basic algorithm. Failing to compare all pairs of points is likely to reveal information about the relative distances of the points that *are* compared. Developing protocols where such revelation can be proven not to disclose information beyond that revealed by simply knowing the outliers is a challenge. Otherwise, completely novel techniques must be developed which do not require *any* pairwise comparison. When there are three or more parties, assuming no collusion, much more efficient solutions that reveal some information can be developed. In the following sections we discuss some of these techniques developed by Vaidya and Clifton[83] for both partitionings of data. While not completely secure, the privacy versus cost tradeoff may be acceptable in some situations. An alternative (and another approach to future work) is demonstrating lower bounds on the complexity of fully secure outlier detection. However, significant work is required to make any of this happen – thus opening a rich area for future work.

Horizontally partitioned data

The most computationally and communication intensive part of the algorithm are the secure comparisons. With horizontally partitioned data, a semi-trusted third party can perform comparisons and return random shares. The two comparing parties just give the values to be compared to the third party to add and compare. As long as the third party does not collude with either of the comparing parties, the comparing parties learn nothing.

The real question is, what is disclosed to the third party? Basically, since the data is horizontally partitioned, the third party has no idea about the respective locations of the two objects. All it can find out is the distance between the two objects. While this is information that is not a part of the result, by itself it is not very significant and allows a tremendous increase in efficiency. Now, the cost of secure comparison reduces to a total of 4 messages (which can be combined for all comparisons performed by the pair, for a constant number of rounds of communication) and insignificant computation cost.

Vertically partitioned data

The simple approach used in horizontal partitioning is not suitable for vertically partitioned data. Since all of the parties share all of the points, partial knowledge about a point does reveal useful information to a party. Instead, one of the remaining parties is chosen to play the part of completely untrusted non-colluding party. With this assumption, a much more efficient secure comparison algorithm has been postulated by Cachin [11] that reveals nothing to the third party. The algorithm is otherwise equivalent, but the cost of the comparisons is reduced substantially.

7.3.8 Summary

This section has presented privacy-preserving solutions for finding distance based outliers in distributed data sets. A significantly important and useful part is the security proofs for the algorithm. The basic proof technique (of simulation) is the same for every secure algorithm under in the Secure Multiparty Computation framework.

8

Future Research - Problems remaining

In a few short years, the field of privacy-preserving data mining has developed a suite of techniques to address many of the standard data mining tasks. These techniques address privacy in a variety of ways; all open new opportunities for data mining in areas where privacy concerns have limited (or may in the future limit) access to data.

What challenges remain? Perhaps the most apparent is adoption; these techniques have not yet seen real-world application. We see two most likely routes to adoption of privacy-preserving data mining technologies, both will demand effort that goes beyond algorithm development. The first route is to develop new markets for data mining: Identify areas where data mining has not even been considered, as the sharing or disclosure of data is inconceivable. While legally protected data on individuals could be the source of such data, a more likely scenario is protecting secrecy of corporate data where collaboration meets competition. The interaction between legally protected data and privacy-preserving data mining technology has not yet been explored by the courts, and until it is a cost-benefit tradeoff between the risk of using data and the rewards from data mining will be difficulty to evaluate. With corporate secrecy, the tradeoffs are clearer and more easily measured. There has been research work moving toward such applications[6, 7], continued progress could well serve as a driver to bring privacy-preserving data mining technologies.

The second route we see as a likely adoption path for this technology is based on the increasing cost of protecting data. While data is rightly viewed as a valuable asset, legislative actions such a EC 95/46 and California SB1386 as well as court cases regarding privacy are driving up the cost of *protecting* that asset. It is rarely the data itself that provides value, instead it is the knowledge that can be gleaned from the data. The case of CardSystems is a clear example; an information security breach resulted in the theft of about 239,000 names and credit card numbers. Of particular importance is that the files stolen were no longer needed to carry out CardSystems' primary task of authorizing and processing credit card transactions, instead the data "consisted of transactions which were not completed for a variety of reasons.

This data was stored for research purposes in order to determine why these transactions did not successfully complete."[71] Although as of press time of this book the future of CardSystems was uncertain, the testimony cited above noted that Visa and American Express had decided to terminate CardSystems as a transactions processor, potentially dealing a fatal blow to the business. While it is not clear that existing privacy-preserving data mining techniques would have enabled the knowledge discovery CardSystems desired from the stolen data, development and use of such technology certainly would have been financially prudent. Such financial drivers could well lead to adoption of new technology as a cost-saving measure as well as a better means of protecting privacy.

Continued development of privacy-preserving data mining techniques will help to address the adoption problem. If techniques already exist that address the needs of data mining users, the cost of adoption will be lowered. One approach to this is through developing a toolkit that can be used to build privacy-preserving data mining solutions. As we have seen, many algorithms for both perturbation and cryptographic approaches reuse a few basic tools: determining original distributions from distorted data, summation, counting of items in common, etc. The challenge is not in implementing these basic building blocks, but in how to *securely* assemble them. The programming challenges are straightforward, but designing an algorithm and proving it secure still demands a level of expertise beyond what can be expected of developers who have the needed domain expertise to build a real-world application. While education (and hopefully this book) will help, frameworks supporting easier privacy proofs are needed.

A second way to speed adoption of privacy-preserving technology is to integrate this technology with existing applications. For example, building perturbation techniques into web survey software could serve as a selling point for that software. Ensuring that the techniques provide for the varieties of analysis that may be needed, without knowing the specific applications in advance, is a challenging issue.

Perhaps the most technically challenging issue is to develop a better understanding of privacy, and how the outcomes of data mining impact privacy. The scarcity of material in Chapter 2 demonstrates the need for more research in this area. Without a clear understanding of how much or little is revealed by the results of a particular data mining process, it is unlikely that privacy-preserving data mining will be fully accepted in highly sensitive fields such as medical research. While it is likely that such fields will use privacy-preserving technology, the real win will come when these fields recognize the technology as sufficiently effective to waive the normal controls put in place when private data is involved. This will be a long process, involving significant work by the research community to fully prove the efficacy of the technology in guaranteeing privacy. Once such guarantees can be made, we may see knowledge discovery that today is inconceivable due to privacy considerations.

References

1. D. Agrawal and C. C. Aggarwal. On the design and quantification of privacy preserving data mining algorithms. In *Proceedings of the Twentieth ACM SIGACT-SIGMOD-SIGART Symposium on Principles of Database Systems*, pages 247–255, Santa Barbara, California, May 21-23 2001. ACM.
2. R. Agrawal, A. Evfimievski, and R. Srikant. Information sharing across private databases. In *Proceedings of ACM SIGMOD International Conference on Management of Data*, San Diego, California, June 9-12 2003.
3. R. Agrawal, T. Imielinski, and A. N. Swami. Mining association rules between sets of items in large databases. In P. Buneman and S. Jajodia, editors, *Proceedings of the 1993 ACM SIGMOD International Conference on Management of Data*, pages 207–216, Washington, D.C., May 26–28 1993.
4. R. Agrawal and R. Srikant. Privacy-preserving data mining. In *Proceedings of the 2000 ACM SIGMOD Conference on Management of Data*, pages 439–450, Dallas, TX, May 14-19 2000. ACM.
5. M. Atallah, E. Bertino, A. Elmagarmid, M. Ibrahim, and V. Verykios. Disclosure limitation of sensitive rules. In *Knowledge and Data Engineering Exchange Workshop (KDEX'99)*, pages 25–32, Chicago, Illinois, Nov. 8 1999.
6. M. J. Atallah, M. Bykova, J. Li, and M. Karahan. Private collaborative forecasting and benchmarking. In *Proc. 2d. ACM Workshop on Privacy in the Electronic Society (WPES)*, Washington, DC, Oct. 28 2004.
7. M. J. Atallah, H. G. Elmongui, V. Deshpande, and L. B. Schwarz. Secure supply-chain protocols. In *IEEE International Conference on E-Commerce*, pages 293–302, Newport Beach, California, June 24-27 2003.
8. D. Barbará, N. Wu, and S. Jajodia. Detecting novel network intrusions using bayes estimators. In *First SIAM International Conference on Data Mining*, Chicago, Illinois, Apr. 5-7 2001.
9. V. Barnett and T. Lewis. *Outliers in Statistical Data*. John Wiley and Sons, 3rd edition, 1994.
10. J. Benaloh. Dense probabilistic encryption. In *Proceedings of the Workshop on Selected Areas of Cryptography*, pages 120–128, Kingston, Ontario, May 1994.
11. C. Cachin. Efficient private bidding and auctions with an oblivious third party. In *Proceedings of the 6th ACM conference on Computer and communications security*, pages 120-127, Kent Ridge Digital Labs, Singapore, 1999. ACM Press.

12. S. Chawla, C. Dwork, F. McSherry, A. Smith, and H. Wee. Toward privacy in public databases. In *Theory of Cryptography Conference*, Cambridge, MA, Feb. 9-12 2005.

13. D. W.-L. Cheung, J. Han, V. Ng, A. W.-C. Fu, and Y. Fu. A fast distributed algorithm for mining association rules. In *Proceedings of the 1996 International Conference on Parallel and Distributed Information Systems (PDIS'96)*, pages 31–42, Miami Beach, Florida, USA, Dec. 1996. IEEE.

14. Y.-T. Chiang, D.-W. Wang, C.-J. Liau, and T. sheng Hsu. Secrecy of two-party secure computation. In *Proceedings of the 19th Annual IFIP WG 11.3 Working Conference on Data and Applications Security*, pages 114–123, Storrs, Connecticut, Aug. 7-10 2005.

15. C. Clifton, M. Kantarcıoğlu, and J. Vaidya. Defining privacy for data mining. In H. Kargupta, A. Joshi, and K. Sivakumar, editors, *National Science Foundation Workshop on Next Generation Data Mining*, pages 126 133, Baltimore, MD, Nov. 1-3 2002.

16. G. Cooper and E. Herskovits. A bayesian method for the induction of probabilistic networks from data. *Machine Learning*, 9(4):309–347, 1992.

17. A. P. Dempster, N. M. Laird, and D. B. Rubin. Maximum likelihood from incomplete data via the EM algorithm (with discussion). *Journal of the Royal Statistical Society*, B 39:1–38, 1977.

18. A. Dobra and S. E. Fienberg. Bounds for cell entries in contingency tables given marginal totals and decomposable graphs. In *Proceedings of the National Academy of Sciences*, number 97, pages 11885–11892, 2000.

19. P. Doyle, J. Lane, J. Theeuwes, and L. Zayatz, editors. *Confidentiality, Disclosure and Data Access: Theory and Practical Applications for Statistical Agencies*. Elsevier, Amsterdam, Holland, 2001.

20. W. Du and M. J. Atallah. Privacy-preserving statistical analysis. In *Proceeding of the 17th Annual Computer Security Applications Conference*, New Orleans, Louisiana, USA, December 10-14 2001.

21. W. Du, Y. S. Han, and S. Chen. Privacy-preserving multivariate statistical analysis: Linear regression and classification. In *2004 SIAM International Conference on Data Mining*, Lake Buena Vista, Florida, Apr. 22-24 2004.

22. W. Du and Z. Zhan. Building decision tree classifier on private data. In C. Clifton and V. Estivill-Castro, editors, *IEEE International Conference on Data Mining Workshop on Privacy, Security, and Data Mining*, volume 14, pages 1–8, Maebashi City, Japan, Dec. 9 2002. Australian Computer Society.

23. W. Du and Z. Zhan. Using randomized response techniques for privacy-preserving data mining. In *The Ninth ACM SIGKDD International Conference on Knowledge Discovery and Data Mining*, pages 505–510, Washington, DC, Aug. 24-27 2003.

24. W. K. Du and C. Clifton, editors. *On Inverse Frequent Set Mining*, Melbourne, Florida, Nov. 19 2003.

25. R. Duda and P. E. Hart. *Pattern Classification and Scene Analysis*. John Wiley & Sons, 1973.

26. Directive 95/46/EC of the European Parliament and of the Council of 24 October 1995 on the protection of individuals with regard to the processing of personal data and on the free movement of such data. *Official Journal of the European Communities*, No I.(281):31–50, Oct. 24 1995.

27. A. Evfimievski, J. Gehrke, and R. Srikant. Limiting privacy breaches in privacy preserving data mining. In *Proceedings of the 22nd ACM SIGACT-SIGMOD-SIGART Symposium on Principles of Database Systems (PODS 2003)*, pages 211–222, San Diego, CA, June 9-12 2003.

28. A. Evfimievski, R. Srikant, R. Agrawal, and J. Gehrke. Privacy preserving mining of association rules. In *The Eighth ACM SIGKDD International Conference on Knowledge Discovery and Data Mining*, pages 217–228, Edmonton, Alberta, Canada, July 23-26 2002.

29. K. J. Ezawa and S. W. Norton. Constructing bayesian networks to predict uncollectible telecommunications accounts. *IEEE Expert*, 11(5):45–51, Oct. 1996.

30. Falcon fraud managerEA. http://www.fairisaac.com/Fairisaac/Solutions/Solutions+by+Function/Falcon+Fraud+Manager.htm, 2005.

31. M. Feingold, M. Corzine, M. Wyden, and M. Nelson. Data Mining Moratorium Act of 2003. U.S. Senate Bill (proposed), Jan. 16 2003.

32. M. Feingold, M. Jeffords, and M. Leahy. Data-mining reporting act of 2003. U.S. Senate Bill (proposed), July 31 2003. Also H.R. 4290.

33. M. J. Freedman, K. Nissim, and B. Pinkas. Efficient private matching and set intersection. In *Eurocrypt 2004*, Interlaken, Switzerland, May 2-6 2004. International Association for Cryptologic Research (IACR).

34. K. Fukunaga. *Introduction to Statistical Pattern Recognition*. Academic Press, San Diego, CA, 1990.

35. B. Goethals, S. Laur, H. Lipmaa, and T. Mielikäinen. On Secure Scalar Product Computation for Privacy-Preserving Data Mining. In C. Park and S. Chee, editors, *The 7th Annual International Conference in Information Security and Cryptology (ICISC 2004)*, volume 3506, pages 104–120, December 2–3, 2004.

36. O. Goldreich. *The Foundations of Cryptography*, volume 2, chapter General Cryptographic Protocols. Cambridge University Press, 2004.

37. O. Goldreich, S. Micali, and A. Wigderson. How to play any mental game - a completeness theorem for protocols with honest majority. In *19th ACM Symposium on the Theory of Computing*, pages 218–229, 1987.

38. D. Goldschlag, M. Reed, and P. Syverson. Onion routing. *Commun. ACM*, 42(2):39–41, Feb. 1999.

39. H. Hamilton, E. Gurak, L. Findlater, and W. Olive. Computer science 831: Knowledge discovery in databases, Feb. 2002.

40. Standard for privacy of individually identifiable health information. *Federal Register*, 67(157):53181–53273, Aug. 14 2002.

41. Standard for privacy of individually identifiable health information. Technical report, U.S. Department of Health and Human Services Office for Civil Rights, Aug. 2003.

42. Z. Huang, W. Du, and B. Chen. Deriving private information from randomized data. In *Proceedings of the 2005 ACM SIGMOD International Conference on Management of Data*, Baltimore, MD, June 13-16 2005.

43. I. Ioannidis, A. Grama, and M. Atallah. A secure protocol for computing dot-products in clustered and distributed environments. In *The 2002 International Conference on Parallel Processing*, Vancouver, British Columbia, Aug. 18-21 2002.

44. G. Jagannathan and R. N. Wright. Privacy-preserving distributed *k*-means clustering over arbitrarily partitioned data. In *Proceedings of the 2005 ACM SIGKDD International Conference on Knowledge Discovery and Data Mining*, pages 593–599, Chicago, IL, Aug. 21-24 2005.

45. M. Kantarcioglu and J. Vaidya. Privacy preserving naive bayes classifier for horizontally partitioned data. In *Workshop on Privacy Preserving Data Mining held in association with The Third IEEE International Conference on Data Mining*, Melbourne, FL, Nov.19-22 2003.

46. M. Kantarcıoğlu and C. Clifton. Privacy-preserving distributed mining of association rules on horizontally partitioned data. In *The ACM SIGMOD Workshop on Research Issues on Data Mining and Knowledge Discovery (DMKD'02)*, pages 24–31, Madison, Wisconsin, June 2 2002.

47. M. Kantarcıoğlu and C. Clifton. Privacy-preserving distributed mining of association rules on horizontally partitioned data. *IEEE Trans. Knowledge Data Eng.*, 16(9):1026–1037, Sept. 2004.

48. M. Kantarcıoğlu, J. Jin, and C. Clifton. When do data mining results violate privacy? In *Proceedings of the 2004 ACM SIGKDD International Conference on Knowledge Discovery and Data Mining*, pages 599–604, Seattle, WA, Aug. 22-25 2004.

49. H. Kargupta, S. Datta, Q. Wang, and K. Sivakumar. On the privacy preserving properties of random data perturbation techniques. In *Proceedings of the Third IEEE International Conference on Data Mining (ICDM'03)*, Melbourne, Florida, Nov. 19-22 2003.

50. A. F. Karr, X. Lin, A. P. Sanil, and J. P. Reiter. Secure regressions on distributed databases. *Journal of Computational and Graphical Statistics*, 14:263 – 279, 2005.

51. E. M. Knorr and R. T. Ng. Algorithms for mining distance-based outliers in large datasets. In *Proceedings of 24th International Conference on Very Large Data Bases (VLDB 1998)*, pages 392–403, New York City, NY, USA, Aug.24-27 1998.

52. E. M. Knorr, R. T. Ng, and V. Tucakov. Distance-based outliers: algorithms and applications. *The VLDB Journal*, 8(3-4):237–253, 2000.

53. A. Lazarevic, A. Ozgur, L. Ertoz, J. Srivastava, and V. Kumar. A comparative study of anomaly detection schemes in network intrusion detection. In *SIAM International Conference on Data Mining (2003)*, San Francisco, California, May 1-3 2003.

54. M. Lewis. Department of defense appropriations act, 2004, July 17 2003. Title VIII section 8120. Enacted as Public Law 108-87.

55. X. Lin, C. Clifton, and M. Zhu. Privacy preserving clustering with distributed EM mixture modeling. *Knowledge and Information Systems*, to appear 2004.

56. Y. Lindell and B. Pinkas. Privacy preserving data mining. In *Advances in Cryptology – CRYPTO 2000*, pages 36–54. Springer-Verlag, Aug. 20-24 2000.

57. Y. Lindell and B. Pinkas. Privacy preserving data mining. *Journal of Cryptology*, 15(3):177–206, 2002.

58. Merriam-webster online dictionary.

59. T. Mitchell. *Machine Learning*. McGraw-Hill Science/Engineering/Math, 1st edition, 1997.

60. R. A. Moore, Jr. Controlled data-swapping techniques for masking public use microdata sets. Statistical Research Division Report Series RR 96-04, U.S. Bureau of the Census, Washington, DC., 1996.

61. D. Naccache and J. Stern. A new public key cryptosystem based on higher residues. In *Proceedings of the 5th ACM conference on Computer and communications security*, pages 59–66, San Francisco, California, United States, 1998. ACM Press.

62. M. Naor and B. Pinkas. Oblivious transfer and polynomial evaluation. In *Proceedings of the Thirty-first Annual ACM Symposium on Theory of Computing*, pages 245–254, Atlanta, Georgia, United States, 1999. ACM Press.

63. J. Neter, M. H. Kutner, W. Wasserman, and C. J. Nachtsheim. *Applied Linear Regression Models*. McGraw-Hill/Irwin, 3rd edition, 1996.

64. Toysmart bankruptcy settlement ensures comsumer privacy protection, Jan. 11 2001.

65. T. Okamoto and S. Uchiyama. A new public-key cryptosystem as secure as factoring. In *Advances in Cryptology - Eurocrypt '98, LNCS 1403*, pages 308–318. Springer-Verlag, 1998.

66. S. R. M. Oliveira and O. R. Zaïane. Achieving privacy preservation when sharing data for clustering. In *Workshop on Secure Data Management in a Connected World (SDM'04) in conjunction with VLDB'2004*, pages 67–82, Toronto, Canada, Aug. 30 2004.

67. S. R. M. Oliveira and O. R. Zaïane. Privacy-preserving clustering by object similarity-based representation and dimensionality reduction transformation. In *Workshop on Privacy and Security Aspects of Data Mining (PSDM'04) in conjunction with the Fourth IEEE International Conference on Data Mining (ICDM'04)*, pages 21–30, Brighton, UK, Nov. 1 2004.

68. P. Paillier. Public key cryptosystems based on composite degree residuosity classes. In *Advances in Cryptology - Eurocrypt '99 Proceedings, LNCS 1592*, pages 223–238. Springer-Verlag, 1999.

69. C. Palmeri. Believe in yourself, believe in the merchandise. *Forbes Magazine*, 160(5):118–124, Sept.8 1997.

70. D. Pelleg and A. Moore. X-means: Extending k-means with efficient estimation of the number of clusters. In *Proceedings of the Seventeenth International Conference on Machine Learning*, pages 727–734, San Francisco, 2000. Morgan Kaufmann.

71. J. M. Perry. Statement of john m. perry, president and ceo, cardsystems solutions, inc. before the united states house of representatives subcommittee on oversight and investigations of the committee on financial services. http://financialservices.house.gov/hearings.asp?formmode= detail&hearing=407&comm=4, July 21 2005.

72. J. R. Quinlan. Induction of decision trees. *Machine Learning*, 1(1):81–106, 1986.

73. S. Ramaswamy, R. Rastogi, and K. Shim. Efficient algorithms for mining outliers from large data sets. In *Proceedings of the 2000 ACM SIGMOD international conference on Management of data*, pages 427–438, Dallas, Texas, United States, 2000. ACM Press.

74. M. K. Reiter and A. D. Rubin. Crowds: Anonymity for Web transactions. *ACM Transactions on Information and System Security*, 1(1):66–92, Nov. 1998.

75. S. J. Rizvi and J. R. Haritsa. Maintaining data privacy in association rule mining. In *Proceedings of 28th International Conference on Very Large Data Bases*, pages 682–693, Hong Kong, Aug. 20-23 2002. VLDB.

76. P. Samarati. Protecting respondent's privacy in microdata release. *IEEE Trans. Knowledge Data Eng.*, 13(6):1010–1027, Nov./Dec. 2001.

77. A. P. Sanil, A. F. Karr, X. Lin, and J. P. Reiter. Privacy preserving regression modelling via distributed computation. In *KDD '04: Proceedings of the tenth ACM SIGKDD international conference on Knowledge discovery and data mining*, pages 677–682, New York, NY, USA, 2004. ACM Press.

78. Y. Saygın, V. S. Verykios, and C. Clifton. Using unknowns to prevent discovery of association rules. *SIGMOD Record*, 30(4):45–54, Dec. 2001.

79. L. Sweeney. k-anonymity: a model for protecting privacy. *International Journal on Uncertainty, Fuzziness and Knowledge-based Systems*, (5):557–570, 2002.

80. J. Vaidya and C. Clifton. Privacy preserving association rule mining in vertically partitioned data. In *The Eighth ACM SIGKDD International Conference on Knowledge Discovery and Data Mining*, pages 639–644, Edmonton, Alberta, Canada, July 23-26 2002.

81. J. Vaidya and C. Clifton. Privacy-preserving *k*-means clustering over vertically partitioned data. In *The Ninth ACM SIGKDD International Conference on Knowledge Discovery and Data Mining*, pages 206–215, Washington, DC, Aug. 24-27 2003.

82. J. Vaidya and C. Clifton. Privacy preserving naïve bayes classifier for vertically partitioned data. In *2004 SIAM International Conference on Data Mining*, pages 522–526, Lake Buena Vista, Florida, Apr. 22-24 2004.

83. J. Vaidya and C. Clifton. Privacy-preserving outlier detection. In *The Fourth IEEE International Conference on Data Mining*, Brighton, UK, Nov. 1-4 2004.

84. J. Vaidya and C. Clifton. Secure set intersection cardinality with application to association rule mining. *Journal of Computer Security*, 13(4), Nov. 2005.

85. V. S. Verykios, A. K. Elmagarmid, B. Elisa, Y. Saygin, and D. Elana. Association rule hiding. *IEEE Trans. Knowledge Data Eng.*, 16(4), 2004.

86. L. Wang, D. Wijesekera, and S. Jajodia. Cardinality-based inference control in data cubes. *Journal of Computer Security*, 12(5):655–692, 2005.

87. S. L. Warner. Randomized response: a survey technique for eliminating evasive answer bias. *Journal of the American Statistical Association*, 60(309):63–69, Mar. 1965.

88. L. Willenborg and T. D. Waal. *Elements of Statistical Disclosure Control*, volume 155 of *Lecture Notes in Statistics*. Springer Verlag, New York, NY, 2001.

89. R. Wright and Z. Yang. Privacy-preserving bayesian network structure computation on distributed heterogeneous data. In *Proceedings of the 10th ACM SIGKDD International Conference on Knowledge Discovery and Data Mining*, Seattle, WA, Aug.22-25 2004.

90. A. C. Yao. How to generate and exchange secrets. In *Proceedings of the 27th IEEE Symposium on Foundations of Computer Science*, pages 162–167. IEEE, 1986.

91. N. Zhang, S. Wang, and W. Zhao. A new scheme on privacy-preserving association rule mining. In *The 8th European Conference on Principles and Practice of Knowledge Discovery in Databases (PKDD 2004)*, Pisa, Italy, Sept. 20-24 2004.

92. Y. Zhu and L. Liu. Optimal randomization for privacy preserving data mining. In *KDD '04: Proceedings of the tenth ACM SIGKDD international conference on Knowledge discovery and data mining*, pages 761–766, New York, NY, USA, 2004. ACM Press.

Index